CISM COURSES AND LECTURES

The series presents lecture notes, monographs, edited works and proceedings in the field of Mechanics, Engineering, Computer Science and Applied Mathematics.
Purpose of the series is to make known in the international scientific and technical community results obtained in some of the activities organized by CISM, the International Centre for Mechanical Sciences.

INTERNATIONAL CENTRE FOR MECHANICAL SCIENCES

COURSES AND LECTURES - No. 472

PLANNING BASED ON DECISION THEORY

EDITED BY

GIACOMO DELLA RICCIA
UNIVERSITY OF UDINE

DIDIER DUBOIS
UNIVERSITY OF TOULOUSE

RUDOLF KRUSE
UNIVERSITY OF MAGDEBURG

HANS-J. LENZ
FREE UNIVERSITY OF BERLIN

Springer-Verlag Wien GmbH

This volume contains 39 illustrations

Originally published by Springer-Verlag Wien New York in 2003

SPIN 10953778

In order to make this volume available as economically and as
rapidly as possible the authors' typescripts have been
reproduced in their original forms. This method unfortunately
has its typographical limitations but it is hoped that they in no
way distract the reader.

ISBN 978-3-211-40756-1   ISBN 978-3-7091-2530-4 (eBook)
DOI 10.1007/978-3-7091-2530-4

PREFACE

This volume contains papers read at the 6th workshop on " Planning based on Decision Theory" Udine (Italy), Sept 26-28, 2002 and prepared for final publication.

As its preceding ones, this workshop took place under the auspices of the International School for the Synthesis of Expert Knowledge (ISSEK) and held in the impressive Palazzo del Torso of the Centre International des Sciences Mécaniques (CISM), Udine.

The workshop was organised jointly by Prof. G. Della Riccia (University of Udine), Prof. D. Dubois (University of Toulouse), Prof. R. Kruse (University of Magdeburg), and Prof. H .- J. Lenz (Free University Berlin). As the workshop was an invitational therefore there was no need for a call for contributed papers. Instead of a CfP the organisers recruited research workers who have had an impact on the main topic of the meeting.

Planning is an area that deals with sequential decision problems. Starting from an initial state, we are interested in finding a sequence of actions in order to achieve a set of predefined goals. Planning goes back to the early 60's with the General Problem Solver, which was the first automated planner published in the literature. Typically, this type of planner assumes a deterministic world that can be handled by unconditional, ever successful actions. Despite its limitations it had a strong impact on follow-up research in Artificial Intelligence. On the contrary Dynamic Programming and Markov Decision Theory, developed in connection with Operational Research consider multi-stage decision making under uncertainty with actions depending on the current state reached. It is of interest to note that Bayesian Belief Network and Influence Diagram Methods have their roots in Dynamic Programming. In recent years, Artificial Intelligence research has focused on planning under uncertainty, bringing the two traditions together, with a stress on partially observable states where successive actions are conditioned on the agent knowledge about the current state. Until now, planning methods were successfully applied in production, logistics, marketing, finance, management, and used in robots, software agents, etc.

In recent years, decision analysis has become again an important technique in business, industry and government. This fact is true due to the strongly increasing influence of communication and co-operation over the Internet. Information is expected to be available in real time, at every site, and disseminated to the right person irrespective of end-user devices, cf. UDDI, SOAP, etc.
Decision analysis provides a rational methodology for decision-making in the face of uncertainty. It enables a decision maker to choose among several alternatives in an optimal fashion, taking chances and risks into the account of the value (utility) of

further information to reduce uncertainty. Decision theory gives a concise framework for making decisions based on models. Its components are a state space forming the range of a set of variables, alternatives or potential actions, and constraints on the decision space a set of consequences of actions, and a preference functional encoding an optimality criterion, involving the costs and value of extra information.

It is evident from above that planning of actions based on decision theory is a hot topic for many disciplines. Seemingly unlimited computing power, networking, integration and collaboration have meanwhile attracted the attention of fields like Machine Learning, Operational Research, Management Science and Computer Science. Software agents of e-commerce, mediators of Information Retrieval Systems and Information Systems are typical new application areas.

Section 1 "Decision Theory.

D. Dubois and Hélène Fargier start on "Qualitative Decision Rules Under Uncertainty". They present more than a survey of qualitative decision theory focused on the discussion of the available decision rules under uncertainty, and their properties. It is shown that the emerging uncertainty theory in qualitative settings is possibility theory rather than probability theory. However these approaches lead to criteria that are sometimes either little decisive due to incomparability, or too adventurous because focusing on the most plausible states, or yet lacking discrimination because or the coarseness of the value scale. Some suggestions to overcome these defects are pointed out.
E. Hüllermeier on "Sequential Decision-Making in Heuristic Search" develops a new approach to case-based decision making within the framework of possibility theory. It leads to a decision-theoretic set-up, which combines the cognitive concepts of belief, preference and similarity. The author focuses on the application of this approach to heuristic search and search-based planning. His main idea is to utilise combinatorial optimisation problems already solved in order to improve future search-based problem solving.
P. Miranda, M. Grabisch and P. Gil apply k-additive measures to decision theory in their paper on "Identification of non-additive measures from sample data". They start from some axiomatic results for general fuzzy measures, and add a further axiom to restrict the fuzzy measure to k-additivity. If data is at hand, different algorithms are proposed to calculate the measure that best fits to the data.
The last paper in this section authored by M. Schaal and H.- J. Lenz is on "Notification Planning with Developing Information States". It is concerned with revision of plans for route guidance due to unpredictable events, which happen while being in action. Intelligent notification keeps track of human users plans and events to be expected in the future. As the information states of the notification system vary in time, best choices for intelligent notification do as well. In order to employ the knowledge about future

information states, timed decision trees are enriched with an explicit notion of time-dependent information states and the impact on notification planning is shown.

Section 2 "Planning, Control and Learning"

P. Traverso deals with incomplete knowledge at planning time and partial observability at execution time in his paper on "The Problem of Planning with Three Sources of Uncertainty". His approach is based on "Planning as Model Checking". A planner called "Model Based Planner" (MB) was built. The author presents some recent results in planning in non-deterministic domains, with partial information available at run time, and for requirements expressed in temporal logic.

In their contribution on "Understanding Control Strategies", I. Bratko and D. Suc explain how a given controller like a crane works and achieves a goal. This corresponds to two settings in which the problems 'Reverse Engineering' and 'Behavioural Cloning' arises. The underlying methodology is qualitative tree learning from numerical data and based on a learning program called QUIN.

In their paper on "Local Sructure Learning in Graphical Models", C. Borgelt and R. Kruse study local structure learning that is based on a decision graph representation of the parameter tables. They explore different schemes to select the conditional distributions to merge and examine the behaviour of evaluation measures. Additionally, they present experimental results on standard test cases, which provide some insight into the interdependence of local and global structure learning.

Section 3 "Application of Planning and Decision Making Theory"

In his paper on " Coordination of the Supply Chain as a Distributed Decision Making Problem", C. Schneeweiss considers the impact of symmetric or asymmetric information and of team / non-team behaviour in supply chain management. As most influential aspects he identifies on a medium-term basis contract models based on co-operative game theory and, on a short-term basis, auctions based on principal agent theory. The impact of an integrated view on the whole supply chain is stressed.

H. Rommelfanger is concerned with fuzzy probabilities and utilities in his paper entitled: " Fuzzy Decision Theory - Intelligent ways for solving real-world decision problems and for saving information costs". He uses fuzzy intervals of the $\varepsilon-\lambda$-type. These special fuzzy sets allow to model vague data in a more flexible way than standard trapezoids. Moreover, the arithmetic operations can be efficiently computed on such imprecise quantities. Various preference orderings on fuzzy intervals are discussed. Based on these definitions the principle of Bernoulli is extended to decision models with fuzzy outcomes. Additional information can be used to improve the prior probabilities.

Moreover, fuzzy probabilities can be used combined with crisp or with fuzzy utilities. New algorithms for calculating the fuzzy expected values are introduced. The author votes for collecting additional information in order to identify a best alternative, however, fairly balanced under cost-benefit considerations.

G. Bamberg and G. Dorfleitner close Section 3 with their paper on "Capital Allocation Under Regret and Kataoka Criteria. They analyse the allocation of a given initial capital between a risk-free and a risky alternative, which means to invest into the stock market or a stock market index. The optimal fraction to invest into the stock market under an expected utility goal has no closed-formula solution. However, the optimal fraction according to the Kataoka regret criterion allows an explicit formula. If studied in the Black/Scholes world, i.e. normally distributed log returns, the allocation problem can be characterised as follows: Under realistic parameter values a_* increases with the length of the planning horizon.

The editors of this volume thank very much all our authors for submitting their papers in time, and Mrs. Angelika Wnuk, Free University Berlin, for her dedicated, diligent secretarial work for the workshop.

We would like to thank the following Institutions for substantial help on various levels:

- The International School for the Synthesis of Expert Knowledge (ISSEK) again for promoting the workshop.
- The University of Udine for administrative support.
- The Centre International des Sciences Mécaniques (CISM) for hosting a group of enthusiastic decision makers and planners..

On behalf of all participants we express our deep gratitude to FONDAZIONE CASSA di RISPARMIO di UDINE e PORDENONE for their financial support.

Giacomo Della Riccia (University of Udine),   May, 2003
Didier Dubois (University of Toulouse),
Rudolf Kruse (University of Magdeburg),
Hans-J. Lenz (Free University Berlin).

CONTENTS

SESSION I

DECISION THEORY

Qualitative Decision Rules Under Uncertainty

Didier Dubois Hélène Fargier

IRIT-CNRS, Université Paul Sabatier, Toulouse, France

Abstract. This paper is a survey of qualitative decision theory focusing on available decision rules under uncertainty, and their properties. It is pointed out that two main approaches exist according to whether degrees of uncertainty and degrees of utility are commensurate (that is, belong to a unique scale) or not. Savage-like axiom systems for both approaches are surveyed. In such a framework, acts are functions from states to results, and decision rules are derived from first principles, bearing on a preference relation on acts. It is shown that the emerging uncertainty theory in qualitative settings is possibility theory rather than probability theory. However these approaches lead to criteria that are either little decisive due to incomparability, or too adventurous because focusing on the most plausible states, or yet lacking discrimination because or the coarseness of the value scale. Some suggestions to overcome these defects are pointed out.

1 Introduction

Traditionally, decision-making under uncertainty (DMU) relies on a probabilistic framework. When modelling a decision maker's rational choice between acts, it is assumed that the uncertainty about the state of the world is described by a probability distribution, and that the ranking of acts is done according to the expected utility of the consequences of these acts. This proposal was made by economists in the 1950's, and justified on an axiomatic basis by Savage (1972) and colleagues. More recently, in Artificial Intelligence, this setting has been applied to problems of planning under uncertainty, and is at the root of the influence diagram methodology for multiple stage decision problems.

However, in parallel to these developments, Artificial Intelligence has witnessed the emergence of a new decision paradigm called qualitative decision theory, where the rationale for choosing among decisions no longer relies on probability theory nor numerical utility functions (Doyle and Thomason, 1999). Motivations for this new proposal are twofold. On the one hand there exists a tradition of symbolic processing of information in Artificial Intelligence, and it is not surprising this tradition should try and stick to symbolic approaches when dealing with decision problems. Formulating decision problems in a symbolic way may be more compatible with a declarative expression of uncertainty and preferences in the setting of some logic-based language (Boutilier, 1994; Thomason, 2000). On the other hand, the emergence of new information technologies like information systems or autonomous robots has generated many new decision problems involving intelligent agents (Brafman and Tennenholtz 1997).

An information system is supposed to help an end-user retrieve information and choose among courses of action, based on a limited knowledge of the user needs. It is not clear that numerical approaches to DMU, that were developed in the framework of economics, are fully adapted to these new problems. Expected utility theory might sound too sophisticated a tool for handling queries of end-users. Numerical utility functions and subjective probabilities presuppose a rather elaborate elicitation process that is worth launching for making strategic decisions that need to be carefully analyzed (see Bouyssou et al. 2000). Users of information systems are not necessarily capable of describing their state of uncertainty by means of a probability distribution, nor may they be willing to quantify their preferences (Boutilier, 1994). This is typical of electronic commerce, or recommender systems for instance. In many cases, it sounds more satisfactory to implement a choice method that is fast, and based on rough information about the user preferences and knowledge. Moreover the expected utility criterion makes full sense for repeated decisions whose successive results accumulate (for instance money in gambling decisions). Some decisions made by end-users are rather one-shot, in the sense that getting a wrong advice one day cannot always be compensated by a good advice the next day. Note that this kind of application often needs multiple-criteria decision-making rather than DMU. However there is a strong similarity between the two problems (Dubois et al. 2002b).

In the case of autonomous robots, conditional plans are often used to monitor the robot behaviour, and the environment of the robots is sometimes only partially observable. The theory of partially observable Markov decision processes leads to highly complex methods, because handling infinite state spaces. A qualitative, finitistic, description of the goals of the robot, and of its knowledge of the environment might lead to more tractable methods (Sabbadin, 2000, for instance). Besides, the expected utility criterion is often adopted because of its mathematical properties (it enables dynamic programming principles to be used). However it is not clear that this criterion is always the most cogent one, for instance in risky environments, where cautious policies should be followed. Anyway, dynamic programming techniques are also compatible with possibility theory (Fargier et al., 1998).

There is a need for qualitative decision rules. However there is no real agreement on what "qualitative" means. Some authors assume incomplete knowledge about classical additive utility models, whereby the utility function is specified via symbolic constraints (Lang, 1996; Bacchus and Grove, 1996 for instance). Others use sets of integers and the like to describe rough probabilities or utilities (Tan and Pearl, 1994). Lehmann (1996) injects some qualitative concepts of negligibility in the classical expected utility framework. However some approaches are genuinely qualitative in the sense that they do not involve any form of quantification. We take it for granted that a qualitative decision theory is one that does not resort to the full expressive power of numbers for the modelling of uncertainty, nor for the representation of utility.

2. Quantitative and qualitative decision rules

A decision problem can be cast in the following framework: consider set S of states (of the world) and a set X of potential consequences of decisions. States encode possible situations, states of affairs, etc. An act is viewed as a mapping f from the state space to the consequence set, namely, in each state s ∈ S, an act f produces a well-defined result f(s) ∈ X. The decision maker must rank acts without knowing what is the current state of the world in a precise way. The consequences of an act can often be ranked in terms of their relative appeal: some consequences are judged better than others. This is often modelled by means of a numerical utility function u which assigns to each consequence x ∈ X a utility value u(x) ∈ R. Classically there are two approaches to modelling the lack of knowledge of the decision maker about the state. The most widely found assumption is that there is a probability distribution p on S. It is either obtained from statistics (this is called decision under risk, Von Neumann and Morgenstern, 1944) or it is a subjective probability supplied by the agent via suitable elicitation methods. Then the most usual decision rule is based on the expected utility criterion:

$$EU(f) = \sum_{s \in S} p(s)u(f(s)). \tag{1}$$

An act f is strictly preferred to act g if and only if EU(f) > EU(g). The expected utility criterion is by far the most commonly used one. This criterion makes sense especially for repeated decisions whose results accumulate. It also clearly presupposes subjective notions like belief and preference to be precisely quantified. It means that, in the expected utility model, the way in which the preference on consequences is numerically encoded will affect the induced preference relation on acts. The model exploits some extra information not contained solely in preference relations on X, namely, the absolute order of magnitude of utility grades. Moreover the same numerical scale is used for utilities and degrees of probability. This is based on the idea that a lottery (involving uncertainty) can be compared to a sure gain or a sure loss (involving utility only) in terms of preference.

Another proposal is the maximin criterion (Arrow and Hurwicz, 1972). It applies when no information about the current state is available, and it ranks acts according to its worst consequence:

$$W^-(f) = \min_{s \in S} u(f(s)). \tag{2}$$

This is the pessimistic criterion, first proposed by Wald (1950). An optimistic counterpart of $W^-(f)$ is obtained by turning minimum into maximum. However, the maximin and maximax criteria do not need numerical utility values. Only a total ordering on consequences is needed. No knowledge about the state of the world is necessary. Clearly this criterion has the major defect of being extremely pessimistic. In practice, it is never used for this reason. Hurwicz has proposed to use a weighted average of $W^-(f)$ and its optimistic counterpart, where the weight bearing on $W^-(f)$ is viewed as a degree of pessimism of the decision maker. Other decision rules have been proposed, especially some that generalize both EU(f) and $W^-(f)$, see (Jaffray, 1989) and (Schmeidler 1989), among others. However, all these extensions again require the quantification of preferences and/or uncertainty.

Qualitative variants of the maximin criterion nevertheless exist. Boutilier (1994) is inspired by preferential inference of nonmonotonic reasoning whereby a proposition p entails another one q by default if q is true in the most normal situations where p is true. He assumes that states of nature are ordered in terms of their relative plausibility using a weak order relation $\geq$ on S. He proposes to choose decisions on the basis of the most plausible states of nature in accordance with the available information, neglecting other states. If the available information is that s $\in$ A, a subset of states, and if A* is the set of maximal elements in A according to the plausibility ordering $\geq$, then the criterion is defined by

$$W_\geq (f) = \min_{s \in A^*} u(f(s)). \tag{3}$$

This approach has been axiomatized by Brafman and Tennenholtz (2000) in terms of conditional policies (rather than acts).

Another refinement of Wald criterion, the possibilistic qualitative criterion is based on a utility function u on X and a possibility distribution π on S (Zadeh, 1978), both mapping on the same totally ordered scale L, with top 1 and bottom 0. The ordinal value $\pi(s)$ represents the relative plausibility of state s. A pessimistic criterion $W^-_\pi(f)$ is proposed of the form (Dubois and Prade, 1988):

$$W^-_\pi(f) = \min_{s \in S} \max(n(\pi(s)), u(f(s))) \tag{4}$$

Here, L is equipped with its involutive order-reversing map n; in particular n(1) = 0, n(0) = 1. So, $n(\pi(s))$ represents the degree of potential surprise (Shackle, 1961) caused by realizing that the state of the world is s. In particular, $n(\pi(s)) = 1$ for impossible states. The value of $W^-_\pi(f)$ is small as soon as there exists a highly plausible state $(n(\pi(s)) = 0)$ with low utility value. This criterion is actually a *prioritized* extension of the Wald maximin criterion $W^-(f)$. The latter is recovered if $\pi(s) = 1$ for all s $\in$ S. The decisions are again made according to the merits of acts in their worst consequences, now restricted to the most plausible states. But the set of most plausible states (S* = {s, $\pi(s) \geq n(W^-_\pi(f))$}) now depends on the act itself. It is defined by the compromise between belief and utility expressed in the min-max expression. However, contrary to the other qualitative criteria, the possibilistic qualitative criterion presupposes that degrees of utility u(f(s)) and possibility $\pi(s)$ share the same scale and can be compared.

The optimistic counterpart to this criterion (Zadeh, 1978) is:

$$W^+_\pi(f) = \max_{s \in S} \min(\pi(s)), u(f(s))) \tag{5}$$

The optimistic criterion has been first proposed by Yager (1979) and the pessimistic criterion by Whalen (1984), and also used by Dubois (1987), Buckley (1988), and Inuiguchi et al. (1989). They have been used for a long time in fuzzy information processing for the purpose of triggering fuzzy rules in expert systems (Cayrol et al., 1982) and flexible querying of an incomplete information database (Dubois et al., 1988). More recently, they were used in scheduling under flexible constraints and uncertain task durations, when minimizing the risk of delayed jobs (Dubois et al, 1995).

These optimistic and pessimistic possibilistic criteria are actually particular cases of a more general criterion based on the Sugeno integral (a qualitative counterpart to Choquet integral (See Grabisch et al., 2000)), one expression of which can be written as follows:

$$S_\gamma(f) = \max_{A \subseteq S} \min(\gamma(A), \min_{s \in A} u(f(s))),\qquad(6)$$

where $\gamma(A)$ is the degree of confidence in event A, and γ is a set-function that reflects the decision-maker attitude in front of uncertainty. This expression achieves a trade-off between the degrees of confidence in events and their worst consequence when they occur. If the set of states is rearranged in decreasing order via f in such a way that $u(f(s_1)) \geq ... \geq u(f(s_n))$, then denoting $A_i = \{s_1,..., s_i\}$, it turns out that $S_\gamma(f)$ is the median of the set $\{u(f(s_1)),..., u(f(s_n))\}$ $\cup \{\gamma(A_1),..., \gamma(A_{n-1})\}$.

The restriction to the most plausible states, as implemented by the above criteria, makes them more realistic than the maximin criterion, but it still yields a very coarse ranking of acts (there are no more classes than the number of elements in the finite scale L). The above qualitative criteria do not use all the available information to discriminate among acts. Especially an act f can be ranked equally with another act g even if f is at least as good as g in all states and better in some states (including most plausible ones). This defect cannot be found with the expected utility model. Lehmann (2001) axiomatizes a refinement of the maximin criterion whereby ties between equivalent worst states are broken by considering their respective likelihoods. This decision rule takes the form of an expected utility criterion with qualitative (infinitesimal) utility levels. An axiomatization is carried out in the Von Neumann-Morgenstern style.

The lack of discrimination of the maximin rule was actually addressed a long time ago by Cohen and Jaffray (1980) who improve it by comparing acts on the basis of their worst consequences of *distinct* merits, i.e. one considers only the set $D(f, g) = \{s, u(f(s)) \neq u(g(s))\}$ when performing a minimization. Denoting by $f >_D g$ the strict preference between acts,

$$f >_D g \text{ iff } \min_{s \in D(f, g)} u(f(s)) > \min_{s \in D(f, g)} u(f(s))\qquad(7)$$

This refined rule always rates an act f better than another act g whenever f is at least as good as g in all states and better in some states (strict compatibility with Pareto-dominance). However, only a partial ordering of acts is then obtained. This last decision rule is actually no longer based on a preference functional. It has been independently proposed by Fargier and Schiex (1993) and used in fuzzy constraint satisfaction problems (Dubois and Fortemps, 1999) under the name " discrimin ordering ".

3. An ordinal decision rule without commensurateness.

Many of the above decision rules presuppose that utility functions and uncertainty functions share the same range, so that it makes sense to write $\min(\pi(s)), u(f(s)))$, for instance. In contrast, one may look for a natural decision rule that computes a preference relation on acts from a purely symbolic perspective, no longer assuming that utility and partial belief are commensurate, that is, share the same totally ordered scale (Dubois Fargier and Prade, 1997). The decision-maker then only supplies a confidence relation between events and a preference ordering on consequences.

In the most realistic model, a confidence relation on the set of events is an irreflexive and transitive relation $>_L$ on 2^S, and a non-trivial one ($S >_L \varnothing$), faithful to deductive inference. A $>_L B$ means that A is more likely than B. Moreover if $A \subseteq B$ then $A >_L B$ should not hold.

The inclusion-monotony property states that if A implies B, then A cannot be more likely than B. Let us define the weak likelihood relation $\geq_L$ induced from $>_L$ via complementation, and the indifference relation $\sim_L$ as usual: A $\geq_L$ B iff not (B $>_L$ A); A $\sim_L$ B iff A $\geq_L$ B and B $\geq_L$ A.

The preference relation on the set of consequences X is supposed to be a *weak order* (a complete preordering, e.g. Roubens and Vincke, 1985). Namely, $\geq_P$ is a reflexive and transitive relation, and completeness means $x \geq_P y$ or $y \geq_P x$. So, $\forall x, y \in X$, $x \geq_P y$ means that consequence x is not worse than y. The induced strict preference relation is derived as usual: $x >_P y$ if and only if $x \geq_P y$ and not $y \geq_P x$. It is assumed that X has at least two elements x and y s.t. $x >_P y$. The assumptions pertaining to $\geq_P$ are natural in the scope of numerical representations of utility, however we do not require that the weak likelihood relation be a weak order too.

If the likelihood relation on events and the preference relation on consequences are not comparable, a natural way of lifting the pair $(>_L, \geq_P)$ to X^S is as follows: an act f is more promising than an act g if and only if the event formed by the disjunction of states in which f gives better results than g, is more likely than the event formed by the disjunction of states in which g gives results better than f. A state s is more promising for act f than for act g if and only if $f(s) >_P g(s)$.

Let $[f >_P g]$ be an event made of all states where f outperforms g, that is $[f >_P g] = \{s \in S, f(s) >_P g(s)\}$. Accordingly, we define the preference between acts ($\geq$), the corresponding indifference ($\sim$) and strict preference ($>$) relations as follows:

$$f > g \quad \text{if and only if} \quad [f >_P g] >_L [g >_P f];$$

$$f \geq g \quad \text{if and only if} \quad \text{not}(g > f) \text{ i.e. if and only if } [f >_P g] \geq_L [g >_P f];$$

$$f \sim g \quad \text{if and only if} \quad f \geq g \text{ and } g \geq f.$$

This is the Likely Dominance Rule (Dubois et al., 1997, 2002a). It is the first one that comes to mind when information is only available under the form of an ordering of events and an ordering of consequences and *when the preference and uncertainty scales are not comparable*. Events are only compared to events, and consequences to consequences. The properties of the relations $\geq$, $\sim$, and $>$ on X^S will depend on the properties of $\geq_L$ with respect to Boolean connectives. An interesting remark is that if $\geq_L$ is a comparative probability ordering then the strict preference relation $>$ in X^S is not necessarily transitive.

Example 1:

A very classical and simple example of undesirable lack of transitivity is when S = {s_1, s_2, s_3} and X = {x_1, x_2, x_3} with $x_1 >_P x_2 >_P x_3$, and the comparative probability ordering is generated by a uniform probability on S. Suppose three acts f, g, h such that

$f(s_1) = x_1 >_P f(s_2) = x_2 >_P f(s_3) = x_3$,
$g(s_3) = x_1 >_P g(s_1) = x_2 >_P g(s_2) = x_3$,
$h(s_2) = x_1 >_P h(s_3) = x_2 >_P h(s_1) = x_3$.

Then $[f >_P g] = \{s_1, s_2\}$; $[g >_P f] = \{s_3\}$; $[g >_P h] = \{s_1, s_3\}$; $[h >_P g] = \{s_2\}$; $[f >_P h] = \{s_1\}$; $[h >_P f] = \{s_2, s_3\}$.

The likely dominance rule yields $f > g$, $g > h$, $h > f$. Note that the presence of this cycle does not depend on figures of utility that could be attached to consequences insofar as the ordering of utility values is respected for each state. In contrast the ranking of acts induced by expected utility completely depends on the choice of utility values, even if we keep the constraint $u(x_1) > u(x_2) > u(x_3)$. The reader can check that, by symmetry, any of the three linear orders $f > g > h$, $g > h > f$, $h > f > g$ can be obtained by the expected utility criterion, suitably quantifying the utility values of states without changing their preference ranking. Note that the undesirable cycle remains if probabilities $p(s_1) > p(s_2) > p(s_3)$ are attached to states, and the degrees of probability remain close to each other (so that $p(s_2) + p(s_3) > p(s_1)$).

This situation can be viewed as analogue to the Condorcet paradox in social choice, here in the setting of DMU. Indeed the problem of ranking acts can be cast in the setting of a voting problem (See Moulin, 1988, for an introduction to voting methods). Let V be a set of voters, C be a set of candidates and let $\geq_v$ be a relation on C that represents the preference of voter v on the set of candidates. $\geq_v$ is a weak order, by assumption. The decision method consists in constructing a relation R on C that aggregates $\{\geq_v, v \in V\}$ as follows. Let $V(c_1, c_2) = \{v \in V, c_1 >_v c_2\}$ be the set of voters who find c_1 more valuable than c_2, and $|V(c_1, c_2)|$ the cardinality of that set. Then the social preference relation R on C is defined as follows by Condorcet:

$$c_1 \ R \ c_2 \text{ if and only if } |V(c_1, c_2)| > |V(c_2, c_1)|.$$

This is the so-called pairwise majority rule. It is well known that such a relation is often not transitive and may contain cycles. More generally, Arrow (1951) proved that the transitivity of R is impossible under natural requirements on the voting procedure such as independence of irrelevant alternatives, unanimity, and non-dictatorship (i.e., $R \quad \geq_v$ for the same voter v systematically). The probabilistic version of the likely dominance rule is very similar to Condorcet procedure, taking $V = S$, $C = X^S$, and considering for each $s \in S$ the relation R_s on acts such that $\forall f, g \in X^S$: $f \ R_s \ g$ if and only if $f(s) >_p g(s)$. Computing the probability Prob($[f >_p g]$) is a weighted version of $|V(c_1, c_2)|$ with $V = S$, $c_1 = f$, $c_2 = g$, which explains the intransitivity phenomenon. Such weighted extensions of Condorcet procedure are commonly found in multicriteria decision-making (Vincke 1992). However, the likely dominance rule makes sense for any inclusion-monotonic likelihood relation between events and is then much more general than the Condorcet pairwise majority rule even in its weighted version.

Assume now that a decision maker supplies a weak order of states in the form of a possibility distribution π on S and a weak order of consequences $\geq_p$ on X. Let $>_\Pi$ be the induced possibilistic ordering of events (Dubois, 1986). Namely, denote max(A) any (most plausible) state $s \in A$ such that $s \geq_\pi s'$ for any $s' \in A$. Then, define $A \geq_\Pi B$ if and only if max(A) $\geq_\pi$ max(B). The preference on acts in accordance with the likely dominance rule, for any two acts f and g, is : $f > g$ iff $[f >_p g] >_\Pi [g >_p f]$; $f \geq g$ iff $\neg(g > f)$. Then, the undesirable intransitivity of the strict preference vanishes.

Example 1 (continued)

Consider again the 3-state/3-consequence example of Section 3. If a uniform probability is changed into a uniform possibility distribution, then it is easy to check that the likely dominance rule yields: $f \sim g \sim h$. However, if $s_1 >_\pi s_2 >_\pi s_3$ then

$[f >_P g] = \{s_1, s_2\} >_\Pi [g >_P f] = \{s_3\}$;

$[g >_P h] = \{s_1, s_3\} >_\Pi [h >_P g] = \{s_2\}$;

$[f >_P h] = \{s_1\} >_\Pi [h >_P f] = \{s_2, s_3\}$.

So $f > g > h$ follows. It contrasts with the cycles obtained with a probabilistic approach. However the indifference relation between acts is generally not transitive.

Let us study the likely dominance rule induced by a single possibility distribution (and the possibilistic likelihood relation it induces). If the decision maker is ignorant about the state of the world, all states are equipossible, and all events but $\emptyset$.

Conversely, if there is an ordering $s_1, \dots, s_n$ of S such that $\pi(s_1) > \pi(s_2) > \dots > \pi(s_n)$, then for any A, B such that $A \cap B = \emptyset$, $A >_\Pi B$ or $B >_\Pi A$. Hence $\forall f$ g, $f > g$ or $g > f$. Moreover this is a lexicographic ranking:

$$f > g \text{ iff } \exists k \text{ such that } f(s_k) >_P g(s_k) \text{ and } f(s_i) \sim_P g(s_i), \forall i < k.$$

It corresponds to the procedure: check if f is better than g in the most normal state; if yes prefer f; if f and g give equally preferred results in s_1, check in the second most normal state, and so on recursively. This comes down to a lexicographic ranking of vectors $(f(s_1), \dots, f(s_n))$ and $(g(s_1), \dots, g(s_n))$. It is a form of dictatorship by most plausible states, in voting theory terms. It also coincides with Boutilier's criterion, except that ties can be broken by less normal states. More generally any weak order splits S into a well-ordered partition $S_1 \cup S_2 \cup \dots \cup S_m = S$, $S_i \cap S_j = \emptyset$ ($i \neq j$), such that states in each S_i are equally plausible and states in S_i are more plausible than states in S_j, $\forall j > i$. Then, the ordering of events is defined as follows:

$$A >_{L\Pi} B \text{ if and only if } \min\{i: S_i \cap A \cap B^c\} < \min\{i: S_i \cap B \cap A^c\},$$

and the ordering of acts is defined by

$f > g$ iff $\exists k \geq 1$ such that: $\forall s \in S_1 \cup S_2 \cup \dots \cup S_{k-1}$, $f(s) \sim_P g(s)$, $\forall s \in S_k$, $f(s) \geq_P g(s)$ and $\exists s \in S_k$, $f(s) >_P g(s)$

$f \sim g$ iff $\exists k \geq 1$ such that: $\forall s \in S_1 \cup S_2 \cup \dots \cup S_{k-1}$, $f(s) \sim_P g(s)$, $\exists s \in S_k$, s' $\in S_k$, s   s', s.t. $f(s) >_P g(s)$ and $g(s') >_P f(s')$.

This decision criterion is a blending of lexicographic priority and unanimity among states. Informally, the decision maker proceeds as follows: f and g are compared on the set of most normal states (S_1): if f performs better than g in S_1, then f is preferred to g; if there is a disagreement in S_1 about the relative performance of f and g then f and g are not comparable. If f and g have equally preferred consequences in each most normal state then the decision maker considers the set of second most normal states S_2, etc. In a nutshell, it is a prioritized Pareto-dominance relation. Preferred acts are selected by restricting to the most plausible states of the world, and a unanimity rule is used on these maximally plausible states. Ties are broken by lower level oligarchies. So this procedure is similar to Boutilier's decision rule in that it focuses on the most plausible states, but Pareto-dominance is required instead of the maximin rule on them, and ties can be broken by subsets of lower plausibility. This decision rule is cognitively appealing, but it has a limited expressive and decisive power. One may apply the maximin rule in a prioritized way: the improved maximin decision rule can be substituted to unanimity within the likely dominance rule inside the oligarchies of states. It is easy to imagine a counterpart to the likely dominance rule where expected utility applies inside the oligarchies of states (Lehmann, 1996). However reasonable these refined decision rules may look, they need to be formally justified.

4. Axiomatics of qualitative decision theory

A natural question is then whether it is possible to found rational decision-making in a purely qualitative setting, under an act-driven framework a la Savage. The idea of the approach is to extract the decision maker's confidence relation and the decision maker's preference on consequences from the decision maker's preference pattern on acts. Enforcing "rationality" conditions on the way the decision maker should rank acts then determines the kind of uncertainty theory implicitly "used" by the decision maker for representing the available knowledge on states. It also prescribes a decision rule. Moreover, this framework is operationally testable, since choices made by individuals can be observed, and the uncertainty theory at work is determined by these choices.

As seen in sections 2 and 3, two research lines can be followed in agreement with this definition: the relational approach and the absolute approach. Following the relational approach, the decision maker uncertainty is represented by a partial ordering relation among events (expressing relative likelihood), and the utility function is just encoded as another ordering relation between potential consequences of decisions. The advantage is that it is faithful to the kind of elementary information users can directly provide. The other approach, which can be dubbed the absolute approach (Dubois Prade and Sabbadin, 2001), presupposes the existence of a totally ordered scale (typically a finite one) for grading both likelihood and utility. Both approaches lead to an act-driven axiomatization of the qualitative variant of possibility theory (Zadeh, 1978; Dubois and Prade, 1998).

4.1. Savage theory: a refresher

The Savage framework is adapted to our purpose of devising a purely ordinal approach because its starting point is indeed based on relations even if their representation is made on an interval scale. Suppose a decision maker supplies a preference relation $\geq$ over acts f: S $\rightarrow$ X. X^S usually denotes the set of all such mappings. In Savage's approach, any mapping in the set X^S is considered as a possible act (even if it is an imaginary one rather than a feasible one). The first requirement stated by Savage is:

Axiom S1: $(X^S, \geq)$ is a weak order.

This axiom is unavoidable in the scope of utility theory: If acts are ranked according to expected utility, then the preference over acts will be transitive, reflexive, and complete (f $\geq$ g or g $\geq$ f for any f, g). What this axiom also implies, if X and S are finite, is that there exists a totally ordered scale, say L, that can serve to evaluate the worth of acts. Indeed the indifference relation (f $\sim$ g if and only if f $\geq$ g and g $\geq$ f) is an equivalence relation, and the set of equivalence classes, denoted $X^S/\sim$ is totally ordered via the strict preference $>$. If $[f]$ and $[g]$ denote the equivalence classes of f and g, $[f] > [g]$ holds if and only if $f > g$ holds for any representatives of each class. So it is possible to rate acts on $L = X^S/\sim$ and $[f]$ is interpreted as the qualitative utility level of f.

An event is modelled by a subset of states, understood as a disjunction thereof. The set of acts is closed under the following combination involving acts and events. Let $A \subseteq S$ be an event, f and g two acts, and denote by fAg the act such that:

$$fAg(s) = f(s) \text{ if } s \in A, \text{ and } g(s) \text{ if } s \notin A. \tag{8}$$

For instance, f may mean "bypass the city", g means "cross the city", and A represents the presence of a traffic jam in the city. Then S represents descriptions of the state of the road network, and X represents a time scale for the time spent by an agent who drives to his/her working place. fAg then means: bypass the city if there is a traffic jam, and cross the city otherwise. More generally the notation $f_1A_1f_2A_2, ..., A_{n-1}f_nA_n$, where $A_1, ..., A_{n-1}, A_n$ is a partition of S, denotes the act whose result is $f_i(s)$ if $s \in A_i$, $\forall$ i = 1, ...n. fAg is actually short for $fAgA^c$ where A^c is the complement of A.

Savage proposed an axiom that he called the sure-thing principle. It requires that the relative preference between two acts does not depend on states where the acts have the same consequences. In other words, the preference between fAh and gAh does not depend on the choice of h:

Axiom S2 (Sure-thing principle): $\forall A, f, g, h, h', fAh \geq gAh$ iff $fAh' \geq gAh'$,

where "iff" is shorthand for "if and only if". For instance, if you bypass the city (f) rather than cross it (g) in case of a traffic jam (A), this preference does not depend on what you would do in case of fluid traffic (A^c), say, cross the city (h = g), bypass it anyway (h = f) or make a strange decision such as staying at home. Grant et al. (2000) pointed out that the name "sure-thing principle" for this postulate was not fully justified since it is hard to grasp where the sure thing is. Grant et al. propose several expressions of a genuine sure-thing principle, one version they called *the weak sure-thing principle* being as follows:

Axiom WSTP: $fAg \geq g$ and $gAf \geq g$ imply $f \geq g$.

The above property really means that the weak preference of f over g does not depend on whether A occurs or not. It is obvious that WSTP is implied by S1 and S2, since from $fAg \geq g = gAg$ and S2 we derive $f = fAf \geq gAf$ and using transitivity of $\geq$ due to S1, $f \geq g$ follows. The sure-thing principle enables two notions to be simply defined, namely conditional preference and null events.

An act f is said to be weakly preferred to g, conditioned on A if and only if $\forall h$, $fAh \geq gAh$. This is denoted by $(f \geq g)_A$. Clearly, the sure-thing principle enables $(f \geq g)_A$ to hold as soon as $fAh \geq gAh$ for some h. Conditional preference $(f \geq g)_A$ means that f is weakly preferred to g when the state space is restricted to A, regardless of the decision made when A does not occur. Note that $f \geq g$ is short for $(f \geq g)_S$. Moreover $(f \geq g)_\varnothing$ always holds, for any f and g, since it is equivalent to the reflexivity of $\geq$ (i.e. $h \geq h$).

An event A is said to be null if and only if $\forall f, \forall g, (f \geq g)_A$ holds. Any non-empty set of states A on which all acts make no difference is then like the empty set: the reason why all acts make no difference is because this event is considered impossible by the decision maker.

Conditional preference enables the weak sure-thing principle to be expressed like a unanimity principle in the terminology of voting theory, provided that the sure-thing principle holds.

Axiom U: $(f \geq g)_A$ and $(f \geq g)_{A^c}$ implies $f \geq g$ (unanimity)

Note that in the absence of S2 (U) implies (WSTP) but not the converse. The unanimity postulate has been formulated by Lehmann (1996).

Among acts in X^S are *constant acts* such that: $\exists\, x \in X$, $\forall\, s \in S$, $f(s) = x$. They are denoted fx. It seems reasonable to identify the set of constant acts {fx, $x \in X$} and X. The preference $\geq_P$ on X can be induced from $(X^S, \geq)$ as follows:

$$\forall x, y \in X, \ x \geq_P y \text{ if and only if } fx \geq fy. \tag{9}$$

This definition is self-consistent provided that the preference between constant acts is not altered by conditioning. This is the third Savage's postulate:

Axiom S3: $\forall\, A \subseteq S$, A not null, $(fx \geq fy)_A$ if and only if $x \geq_P y$. $\tag{10}$

Clearly, Pareto-dominance should imply weak preference for acts. And indeed under S1, S2, and S3, $f \geq_P g$ implies $f \geq g$.

The preference on acts also induces a likelihood relation among events. For this purpose, it is enough to consider the set of binary acts, of the form fxAfy, which due to (S3) can be denoted xAy, where $x \in X$, $y \in X$, and $x >_P y$. Clearly for fixed $x >_P y$, the set of acts $\{x,y\}^S$ is isomorphic to the set of events 2^S. However the restriction of $(X^S, \geq)$ to $\{x, y\}^S$ may be inconsistent with the restriction to $\{x', y'\}^S$ for other choices of consequences $x' >_P y'$. A relative likelihood $\geq_L$ among events can however be recovered, as suggested by Lehmann (1996):

$$\forall A, B \subseteq S, \ A \geq_L B \text{ if and only if } xAy \geq xBy, \ \forall x, y \in X \text{ such that } x >_P y.$$

In order to get a weak order of events, Savage introduced a new postulate:

Axiom S4: $\forall x, y, x', y' \in X$ s.t. $x >_P y$, $x' >_P y'$, then $xAy \geq xBy$ iff $x'Ay' \geq x'By'$.

Under this property, the choice of $x, y \in X$ with $x >_P y$ does not affect the ordering between events in terms of binary acts, namely: $A \geq_L B$ is short for $\exists\, x >_P y$, $xAy \geq xBy$.

Lastly, Savage assumed that the set X is not trivial:

Axiom S5: X contains at least two elements x, y such that $fx > fy$ (or $x >_P y$).

Under S1-S5, the likelihood relation on events is a comparative probability ordering (see Fishburn(1986)), i.e. satisfies the preadditivity property

$$\text{If } A \cap (B \cup C) = \varnothing, \text{ then } B \geq_L C \text{ iff } A \cup B \geq_L A \cup C \tag{11}.$$

Such relations are induced by probability measures, but the converse is not true. Savage introduces another postulate that enables him to derive the existence (and uniqueness) of a numerical probability measure on S that can represent the likelihood relation $\geq_L$. This axiom reads:

Axiom S6: For any f, g with $f > g$ in X^S and any $x \in X$, there is a partition $\{S_1, ..., S_n\}$ of S such that $\forall$ $i = 1, ...n$, $xS_i f > g$ and $f > xS_i g$.

Under the postulates S1-S6, not only can $\geq_L$ be represented by a numerical probability function but $(X^S, \geq)$ can be represented by the expected utility of acts $u(f) = \int_{s \in S} u(f(s)) dP(s)$ where the utility function u represents the relation $\geq_P$ on X uniquely, up to a linear transformation. However S6 presupposes that the state space S is infinite so that the probability of S_i can be made arbitrary small, thus not altering the relation $f > g$ when x is very bad (so that $xS_i f > g$) or very good (so that $f > xS_i g$). In contrast, we assume that both S and X are finite in this paper, and S6 is trivially violated in such a finite setting. There is to our knowledge no joint representation of subjective probability and expected utility that would assume a purely finite setting for states and consequences.

4.2 The relational approach to decision theory

Under the relational approach described in Dubois et al. (1997, 2002a, 2003), we have tried to lay bare the formal consequences of adopting a purely ordinal point of view on DMU, while retaining as much as possible from Savage's axioms, and especially the sure thing principle which is the cornerstone of the theory. To this end, an axiom of *ordinal invariance*, originally due to Fishburn (1975) in another context, is then added (Fargier and Perny, 1999). This axiom says that what matters for determining the preference between two acts is the relative positions of consequences of acts for each state, not the consequences themselves, nor the positions of these acts relative to other acts. More rigorously, two pairs of acts (f, g) and (f', g') such that $\forall s \in S$, $f(s) \geq_P g(s)$ if and only if $f'(s) \geq_P g'(s)$ are called *statewise order-equivalent*. This is denoted $(f, g) \equiv (f', g')$. It means that in each state consequences of f, g, and of f', g', are rank-ordered likewise. The *Ordinal Invariance* axiom is:

OI: $\forall f, f' g, g' \in X^S$, if $(f, g) \equiv (f', g')$ then ($f \geq g$ iff $f' \geq g'$).

It expresses the purely ordinal nature of the decision criterion. It is easy to check that the likely dominance rule obeys axiom OI. This is obvious noticing that if $(f, g) \equiv (f', g')$ then by definition, $[f >_P g] = \{s, f(s) >_P g(s)\} = [f' >_P g']$. More specifically, under OI, if the weak preference on acts is reflexive and the induced weak preference on consequences is complete, the only possible decision rule is likely dominance, and OI implies the validity of Savage' S2, and S4 axioms.

Adopting axiom OI and sticking to a transitive weak preference on acts lead to problems exemplified in the previous section by the probabilistic variant of the likely dominance rule. Indeed the following result was proved in Dubois et al. (2002a):

Theorem: Suppose that $(X^S, \geq)$ is a weak order satisfying axiom OI, that S and X have at least three elements, then for the likelihood relation $>_L$ (induced by S4) there is a permutation of the *non-null* elements of S, such that $s_1 >_L s_2 >_L \cdots >_L s_{n-1} \geq_L s_n >_L \varnothing$ and $\forall i = 1, ...n - 2$, $s_i >_L \{s_{i+1}, ..., s_n\}$

If X only has two consequences of distinct values, then such a trivialization is avoided. Nevertheless in the general case where X has more than two elements, the Ordinal invariance axiom forbids a Savagean decision maker to believe that there are two equally likely states of the world, each of which being more likely than a third state. This is clearly not acceptable in practice. If we analyze the reason why this phenomenon occurs, it is found that axiom S1 plays the crucial role, as do all as we wish to keep the sure-thing principle (Dubois et al., 2002a). S1 assumes the full transitivity of the likelihood relation $\geq_L$. Dropping an the transitivity of $\geq_L$ suppresses the unnatural restriction of an almost total ordering of states: we are led to a weak form of S1:

WS1: $(X^S, >)$ is a transitive, irreflexive, partially ordered set.

Dropping the transitivity of $\geq$ cancels some useful consequences of the sure-thing principle under S1, which are nevertheless consistent with the likely dominance rule. For instance, WSTP (or equivalently, the unanimity axiom U) will not follow from the relaxed framework. We must add it to get it.

As a consequence, if one insists on sticking to a purely ordinal view of DMU, we come up to the framework defined by axioms WS1, WSTP (or U), S3, S5, and OI. The likelihood relation induced by S4 is in agreement with classical deduction:

$$\text{If } B >_L A \text{ then } B \cup C >_L A \text{ and } B >_L A \cap C \tag{12}$$

The null events are then all subsets of a subset N of null states. Moreover, if X has more than two elements, it satisfies the following *strongly* non-probabilistic property (Dubois and Prade, 1995a): for any three pairwise disjoint non-null events A, B, C,

$$B \cup C >_L A \text{ and } A \cup C >_L B \text{ imply } C >_L A \cup B. \tag{13}$$

$B >_L A$ then really means that B is much more likely than A, because B will still be more likely than any disjunction of events nor more likely than A. This likelihood relation can always be represented by a family of possibility relations. Namely, there is a family F of possibility relations (Lewis, 1973; Dubois 1986) on S and a weak order relation $\geq_P$ on X such that the preference relation on acts is defined by

$$f > g \text{ iff } \forall >_\Pi \in F, [f >_P g] >_\Pi [g >_P f]. \tag{14}$$

This ordinal Savagean framework actually leads to a representation of uncertainty that is at work in the nonmonotonic logic system of Kraus, Lehmann and Magidor (1990), as shown by Friedman and Halpern (1996) who also study property (13).

A more general setting starting from a reflexive weak preference relation on acts is used in Dubois et al. (2002b, 2003). In this framework S3 is replaced by a monotonicity axiom on both sides, that is implied by Savage's framework, namely for any event A:

$$\text{If } \forall s, h(s) >_P f(s) \text{ and } f \geq g \text{ then } fAh \geq g;$$

$$\text{If } \forall s, g(s) >_P h(s) \text{ and } f \geq g \text{ then } f \geq gAh.$$

An additional axiom of anonymity, stating that exchanging the consequences of two equally plausible states does not alter the decision maker's preference pattern, implies that the likelihood relation can be represented by a single possibility relation.

The restricted family of decision rules met by the purely relational approach to the decision problem under uncertainty only reflects the situation faced in voting theories where natural axioms lead to impossibility theorems (see for instance, Arrow, 1951; Sen, 1986). This kind of impediment was already pointed out by Doyle and Wellman (1991) for preference-based default theories. These results question the very possibility of a purely ordinal solution to this problem, in the framework of transitive and complete preference relations on acts.

The likely dominance rule lacks discrimination, not because of indifference between acts, but because of incomparability. Actually, it may be possible to weaken axiom OI while avoiding the notion of certainty equivalent of an uncertain act. It must be stressed that OI requires more than the simple ordinal nature of preference and uncertainty (i.e. more than separate ordinal scales for each of them). Condition OI also involves a condition of independence with respect to irrelevant alternatives (in the sense of Arrow 1951). It says that the preference f > g only depends on the relative positions of quantities f(s) and g(s) on the preference scale. This unnecessary part of the condition could be cancelled within the proposed framework, thus leaving room for a new family of rules not considered in this paper, for instance involving a third act or some prescribed consequence considered as an aspiration level.

4. 3 Qualitative decision rules under commensurateness.

Let us now consider the absolute qualitative criteria (4), (5), (6), based on Sugeno integral in the scope of Savage theory. Clearly, they satisfy S1. However the sure thing principle can be severely violated by Sugeno integral. It is easy to show that there may exist f, g, h, h' such that fAh > gAh while gAh' > fAh'. It is enough to consider binary acts (events) and notice that, generally if A is disjoint from $B \cup C$, nothing forbids, for a fuzzy measure γ, to satisfy $\gamma(B) > \gamma(C)$ along with $\gamma(A \cup C) > \gamma(A \cup B)$ (for instance, belief functions are such). The possibilistic criteria (4), (5) violate the sure-thing principle to a lesser extent since:

$$\forall A \subseteq S, \forall f, g, h, h', W^-_\pi(fAh) > W^-_\pi(gAh) \text{ implies } W^-_\pi(fAh') \geq W^-_\pi(gAh')$$

And likewise for W^+_π. Moreover, only one part of S3 holds, for Sugeno integrals. The obtained ranking of acts satisfies the following

Axiom WS3: $\forall A \subseteq S, \forall f, x \geq_P y$ implies $xAf \geq yAf$.

Besides, axiom S4 is violated by Sugeno integrals, but to some extent only. Namely, $\forall x, y, x', y' \in X$ s.t. $x >_P y, x' >_P y'$:

$$S_\gamma(xAy) > S_\gamma(xBy) \text{ implies } S_\gamma(x'Ay') \geq S_\gamma(x'By') \qquad (15)$$

which forbids preference reversals when changing the pair of consequences used to model events A and B. Moreover the strict preference is maintained if the pair of consequences is changed into more extreme ones:

$$\text{If } x' >_P x >_P y >_P y' \text{ then } S_\gamma(xAy) > S_\gamma(xBy) \text{ implies } S_\gamma(x'Ay') > S_\gamma(x'By') \qquad (16)$$

Sugeno integral and its possibilistic specializations are weakly Pareto-monotonic since $\forall f$, $f \geq_P g$ implies $S_\gamma(f) \geq S_\gamma(g)$, but one may have $f(s) >_P g(s)$ for some state s, while $S_\gamma(f) = S_\gamma(g)$.

This is the so-called drowning effect, which also appears in the violations of S4. It is because some states are neglected when comparing acts.

The basic properties of Sugeno integrals exploit disjunctive and conjunctive combinations of acts. Namely, given a preference relation $(X^S, \geq)$, and two acts f and g, define f∧g and f∨g as follows

$$f \wedge g \ (s) = f(s) \text{ if } g(s) \geq_P f(s), \text{ and } g(s) \text{ otherwise}$$

$$f \vee g \ (s) = f(s) \text{ if } f(s) \geq_P g(s), \text{ and } f(s) \text{ otherwise}$$

Act f∧g always produces the worst consequences of f and g in each state, while f∨g always makes the best of them. They are union and intersection of fuzzy sets viewed as acts. Obviously $S_\gamma(f \wedge g) \leq \min(S_\gamma(f), S_\gamma(g))$ and $S_\gamma(f \vee g) \geq \max(S_\gamma(f), S_\gamma(g))$ from weak Pareto monotonicity. These properties hold with equality whenever f or g is a constant act. These properties are in fact characteristic of Sugeno integrals for monotonic aggregation operators (e.g. Dubois Marichal et al, 2001). Actually, these properties can be expressed by means of axioms, called restricted conjunctive and disjunctive dominance (RCD and RDD) on the preference structure $(X^S, \geq)$:

Axiom RCD: if f is a constant act, f > h and g > h imply f∧g > h

Axiom RDD: if f is a constant act, h > f and h > g imply h > f∨g

For instance, RCD means that limiting from above the potential utility values of an act g, that is better than another one h, to a constant value that is better than the utility of act h, still yields an act better than h. This is in contradiction with expected utility theory. Indeed, suppose g is a lottery where you win 1000 euros against nothing with equal chances. Suppose the certainty equivalent of this lottery is 400 euros, received for sure, and h is the fact of receiving 390 euros for sure. Now, it is likely that, if f represents the certainty-equivalent of g, f∧g will be felt strictly less attractive than h, as the former means you win 400 euros against nothing with equal chances. Axiom RCD implies that such a lottery should ever be preferred to receiving 400 − ε euros for sure, for arbitrary small values of ε. This axiom is thus strongly counterintuitive in the context of economic theory, with a continuous consequence set X. However the range of validity of qualitative decision theory is precisely when both X and S are finite. Two presuppositions actually underlie axiom RCD (and similar ones for RDD)

1) There is no compensation effect in the decision process: in case of equal chances, winning 1000 euros cannot compensate the possibility of not earning anything. It fits with the case of one-shot decisions where the notion of certainty equivalent can never materialize: you can only get 1000 euros or get nothing if you just play once. You cannot get 400 euros. The latter can only be obtained in the average, by playing several times.

2) There is a big step between one level $\lambda_i \in L$ in the qualitative value scale and the next one λ_{i+1} with $L = \{1 = \lambda_1 > ... > \lambda_n = 0\}$. The preference pattern f > h always means that f is *significantly* preferred to h so that the preference level of f ∧g can never get very close to that of h when g > h. The counterexample above is obtained by precisely moving these two

preference levels very close to each other so that f ∧g can become less attractive than the sure gain h. Level λ_{i+1} is in some sense negligible in front of λ_i.

Sugeno integral can be axiomatized in the style of Savage (Dubois et al. 1998). Namely, if the preference structure $(X^S, \geq)$ satisfies S1, WS3, S5, RCD and RDD, then there a finite chain of preference levels L, an L-valued possibility set-function γ, and an L-valued utility function on the set of consequences X, such that the preference relation on acts is defined by comparing acts f and g by means of $S_\gamma(f)$ and $S_\gamma(g)$.

Namely, S1, WS3, and S5 imply Pareto-monotonicity. In the representation method, L is the quotient set $X^S/\sim$, the utility value u(x) is the equivalence class of the constant act fx, the degree of likelihood γ(A) is the equivalence class of the binary act 1A0, having extreme consequences.

It is easy to check that the equalities $W^-_\pi(f \land g) = \min(W^-_\pi(f), W^-_\pi(g))$ and $W^+_\pi(f \lor g) = \max(W^+_\pi(f), W^+_\pi(g))$ hold with any two acts f and g, for the pessimistic and the optimistic possibilistic preference functionals respectively. The criterion $W^-_\pi(f)$ can thus be axiomatized by strengthening the axioms RCD as follows:

Axiom CD: ∀f, g, h, f > h and g > h imply f ∧g > h   (conjunctive dominance)

Together with S1, WS3, RDD and S5, CD implies that the set-function γ is a necessity measure and so, $S_\gamma(f) = W^-_\pi(f)$ for some possibility distribution π.

Similarly, the criterion $W^+_\pi(f)$ can be axiomatized by strengthening the axioms RDD as follows

Axiom DD: ∀f, g, h, h > f and h > g imply h > f ∨g   (disjunctive dominance)

Together with S1, WS3, RCD and S5, DD implies that the set-function γ is a possibility measure and so, $S_\gamma(f) = W^+_\pi(f)$ for some possibility distribution π.

In order to figure out why axiom CD leads to a pessimistic criterion, Dubois Prade and Sabbadin (2001) have noticed that it can be equivalently replaced by the following property:

$$\forall A \subseteq S, \forall f, g, fAg > g \text{ implies } g \geq gAf \tag{17}$$

This property can be explained as follows: if changing g into f when A occurs results in a better act, the decision maker has enough confidence in event A to consider that improving the results on A is worth trying. But, in this case, there is less confidence on the complement A^c than in A, and any possible improvement of g when A^c occurs is neglected. Alternatively, the reason why fAg > g holds may be that the consequences of g when A occurs are very bad and the occurrence of A is not unlikely enough to neglect them, while the consequences of g when A^c occurs are acceptable. Then suppose that consequences of f when A occurs are acceptable as well. Then fAg > g. But act gAf remains undesirable because, regardless of the consequences of f when A^c occurs are acceptable, act gAf still possesses plausibly bad consequences when A occurs. So, g ≥ gAf. For instance, g means losing (A) or winning (A^c) 10,000 euros with equal chances according to whether A occurs or not, and f means winning either nothing (A) or 20,000 euros (A^c) conditioned on the same event. Then fAg is clearly safer than g as there is no risk of losing money. However, if (17) holds, then the chance of winning much more money (20,000 euros) by choosing act gAf is neglected because there is still a good chance to lose 10,000 euros with this lottery. Such a behaviour is clearly cautious. The optimistic counterpart to (17) that can serve as a substitute to axiom DD for the representation of criterion W^+_π is:

$$\forall A \subseteq S, \forall f, g, g > fAg \text{ implies } gAf \geq g \qquad (18)$$

5 Toward more efficient qualitative decision rules

The absolute approach to qualitative decision criteria is simple (especially in the case of possibility theory). It looks more realistic and flexible than the likely dominance rule, but has some shortcomings. First one has to accept the commensurateness between utility and degrees of likelihood. It assumes the existence of a common scale for grading uncertainty and preference. It can be questioned, although it is already taken for granted in classical decision theory (via the notion of certainty equivalent of an uncertain event). It is already implicit in Savage approach, and looks acceptable for decision under uncertainty (but more debatable in social choice). Of course, the acts are then totally preordered.

More importantly, absolute qualitative criteria lack discrimination due to many indifferent acts. They are consistent with Pareto dominance only in the wide sense. The Sure Thing principle is violated (even if not drastically for possibilistic criteria). The obtained ranking of decisions is bound to be coarse since there cannot be more classes of preference-equivalent decisions than levels in the finite scale used. This decision rule lacks discrimination power. The above possibilistic decision rule and the maximin rule can consider two acts as indifferent even if one Pareto-dominates the other.

Giang and Shenoi (2000, 2001) have tried to obviate the need for making assumptions on the pessimistic or optimistic attitude of the decision-maker and improve the discrimination power in the absolute qualitative setting by using, as a utility scale, a totally ordered set of possibility measures on a two element set $\{0, 1\}$ containing the values of the best and the worst consequences. Each such possibility distribution represents a qualitative lottery. Let $L\Pi$ = $\{(a, b), \max(a, b) = 1, a, b \in L\}$, with $\pi(0) = a$, $\pi(1) = b$. It can be viewed as a bipolar scale ordered by the following complete preordering relation:

$$(a, b) \leq (c, d) \text{ iff either } a = c = 1 \text{ and } b \leq d$$

$$\text{or } b = d = 1 \text{ and } a \geq c$$

$$\text{or } a = 1 \text{ and } d = 1.$$

The bottom of this utility scale is $(1, 0)$, its top is $(0, 1)$ and its neutral point $(1, 1)$ means « indifferent ». The canonical example of such a scale is the set of pairs $(\Pi(A^c), \Pi(A))$ of degrees of possibility for event A = " getting the best consequence ", and its complement. When $(\Pi(A^c), \Pi(A)) < (\Pi(B^c), \Pi(B))$, it means that B is more likely (certain or plausible) than A (because it is equivalent to $\Pi(A) < \Pi(B)$ or $N(A) < N(B)$. Each consequence x has a utility value (α, β) in $L\Pi$. The proposed preference functional maps acts, viewed as n-tuples f = $((\alpha_1, \beta_1),..., (\alpha_n, \beta_n))$ of values of $L\Pi$, to $L\Pi$ itself using possibility weights $(\pi_1,..., \pi_n)$ of states, such that $\max_{i = 1, ...,n} \pi_i = 1$. The utility of an act f, is computed as the pair

$$W^{GS}(f) = (\max_{i = 1, ...,n} \min (\pi_i, \alpha_i), \max_{i = 1, ...,n} \min (\pi_i, \beta_i)) \in L\Pi. \qquad (19)$$

This form results from simple very natural axioms on possibilistic lotteries, which are counterparts to the Von Neumann and Morgenstern axioms: complete preorder of acts, increasingness in the wide sense according to the ordering in $L\Pi$, substitutability of indifferent lotteries, and the assumption that any consequence of an act is valued on $L\Pi$. Yet, this criterion has a major drawback: Whenever there are two states i and j are such that $\alpha_i = 1$ and $\beta_j = 1$ (respectively a bad or neutral, and a good or neutral state) and these states have maximal possibility $\pi_i = \pi_j = 1$, then u(f) = (1, 1) results, expressing indifference. This limited expressiveness seems to be unavoidable when using finite bipolar scales (Grabisch et al, 2002).

The drowning effect of the possibilistic criteria can be fixed, in the face of total ignorance as done by Cohen and Jaffray on the minimax criterion (7) as pointed out in section 2 (" discrimin " partial order). This criterion can be further refined by the so-called leximin ordering (Moulin, 1988): The idea is to reorder utility vectors $(u(f(s_1)), \ldots u(f(s_n)))$ by non-decreasing values as $(u(f(s_{\sigma(1)})), \ldots u(f(s_{\sigma(n)})))$, where σ is a permutation such that $u(f(s_{\sigma(1)})) \leq u(f(s_{\sigma(2)})) \leq \ldots \leq u(f(s_{\sigma(n)}))$. Let τ be the corresponding permutation for an act g. Define the leximin criterion $>_{leximin}$ as:

$f >_{leximin} g$ iff $\exists\, k \leq n$ such that $\forall\, i < k,\ u(f(s_{\sigma(i)})) = u(g(s_{\tau(i)}))$ and $u(f(s_{\sigma(k)})) > u(g(s_{\tau(k)}))$.

The two possible decisions are indifferent if and only if the corresponding reordered vectors are the same. The leximin-ordering is a refinement of the discrimin ordering, hence of both the Pareto-ordering and the maximin-ordering (Dubois, Fargier and Prade, 1996): $f >_D g$ implies $f >_{leximin} g$. Leximin optimal decisions are always discrimin maximal decisions, and thus indeed min-optimal and Pareto-maximal: $>_{leximin}$ is the most selective among these preference relations. Converse implications are not verified. The leximin ordering can discriminate more than any symmetric aggregation function, since when, e.g. the sum of the $u(f(s_{\sigma(i)}))$'s equals the sum of the $u(g(s_{\tau(i)}))$'s, it does not mean that the reordered vectors are the same.

Interestingly, the qualitative leximin rule can be simulated by means of a sum of utilities provided that the levels in the qualitative utility scale are mapped to values sufficiently far from one another on a numerical scale. Consider a finite utility scale $L = \{\lambda_0 < \ldots < \lambda_m\}$. Let $\alpha_i = u(f(s_i))$, and $\beta_i = u(g(s_i))$. Consider an increasing mapping ψ from L to the reals whereby $a_i = \psi(\alpha_i)$ and $b_i = \psi(\beta_i)$. It is possible to define this mapping in such a way that

$$\min_{i = 1, \ldots n} \alpha_i > \min_{i = 1, \ldots n} \beta_i \text{ implies } \sum_{i = 1, \ldots n} a_i > \sum_{i = 1, \ldots n} b_i \qquad (20)$$

To see it, assume that $\min_{i = 1, \ldots n} \alpha_i = \alpha_k$ and $\min_{i = 1, \ldots n} \beta_i = \beta_j$; so we need that

$$\alpha_k > \beta_j \text{ implies } \sum_{i = 1, \ldots n} a_i > \sum_{i = 1, \ldots n} b_i$$

The worst case is when $a_k = \psi(\lambda_p)$ $b_j = \psi(\lambda_{p-1})$, $a_i = a_k$ for all i, and $b_i = b_m$ for all i   j.
So a sufficient condition for (20) is that

$$m\psi(\lambda_p) > \psi(\lambda_{p-1}) + (m-1)\psi(\lambda_m). \qquad (21)$$

Assume $\psi(\lambda_m) = 1$ and $\psi(\lambda_0) = 0$, and denote $\psi(\lambda_p) = e_p \in [0, 1]$. It is easy to show that (21) induces the following constraints on numerical utility levels e_p:

$$e_1 > (m - 1)/m;\ e_2 > (m^2 - 1)/\ m^2;\ \ldots e_p > (m^p - 1)/\ m^p \ldots \text{until } p = m - 1.$$

Moreover it can be checked that the leximin ordering comes down to applying the Bernoulli criterion with respect to a concave utility function $\psi \circ u$

$$f >_{leximin} g \text{ iff } \sum_{i = 1, \ldots, n} \psi(u(l(s_i))) > \sum_{i = 1, \ldots, n} \psi(u(g(\sigma_i)))$$

The optimistic maximax criterion can be refined similarly by a leximax ordering which can also be simulated by an additive criterion. Consider an increasing mapping ϕ from L to [0, 1], $\phi(\lambda_m) = 1$ and $\phi(\lambda_0) = 0$, where $a_i = \phi(\alpha_i)$ and $b_i = \phi(\beta_i)$. It is possible to define this mapping in such a way that

$$\max_{i = 1, \ldots, n} \alpha_i > \max_{i = 1, \ldots, n} \beta_i \text{ implies } \sum_{i = 1, \ldots, n} a_i > \sum_{i = 1, \ldots, n} b_i \qquad (22)$$

To see it, assume that $\max_{i = 1, \ldots, n} \alpha_i = \alpha_k$ and $\max_{i = 1, \ldots, n} \beta_i = \beta_j$; so we need that

$$\alpha_k > \beta_j \text{ implies } \sum_{i = 1, \ldots, n} a_i > \sum_{i = 1, \ldots, n} b_i \qquad (23)$$

The worst case is when $a_k = \psi(\lambda_p)$; $b_j = \psi(\lambda_{p-1})$, $a_i = 0$ for all i ≠ k, and $b_i = b_j$ for all i. So a sufficient condition for (23) is that $\psi(\lambda_p) > m\psi(\lambda_{p-1})$. It induces the following constraints on numerical utility levels e_p:

$$e_{m-1} < 1/m;\ e_{m-2} < 1/\ m^2;\ \ldots e_p < 1/\ m^p \ldots \text{down until } p = 1.$$

Moreover it can be checked that the leximax ordering comes down to applying the Bernoulli criterion with respect to a convex utility function $\phi \circ u$

$$f >_{leximax} g \text{ iff } \sum_{i = 1, \ldots, n} \phi(u(f(s_i))) > \sum_{i = 1, \ldots, n} \phi(u(g(s_i))). \qquad (24)$$

Interestingly, these qualitative pessimistic and optimistic criteria under ignorance are refined by means of the expected value of a risk-averse and risk-prone utility function respectively with respect to a uniform probability, as can be seen by plotting L against numerical values $\psi(L)$ and $\phi(L)$. These results have been recently extended to possibilistic qualitative criteria $W^-_\pi(f)$ and $W^+_\pi(f)$ by Fargier and Sabbadin (2003). They show how to refine possibilistic utilities by means of weighted averages, thus recovering Savage five first axioms. It corresponds to generalizations of leximin and leximax to prioritized minimum and maximum aggregations, thus bridging the gap between possibilistic criteria and classical decision theory.

The underlying uncertainty representation is then probabilistic. Consider a subset A of states and a finitely scaled possibility distribution π. Let v_A be the vector $(a_1, \ldots, a_n)$ such that, denoting $S = \{s_1, s_2, \ldots, s_n\}$:

$$a_i = \pi(s_i) \text{ if } s_i \in A,$$

$$a_i = 0 \text{ otherwise.}$$

It is possible to assign a vector v_A to each event A. Define a weak ordering among events by

$$A \geq_{\Pi Lex} B \Box\ \ v_A \geq_{leximax} v_B \qquad (25)$$

This relation is called "possibilistic lexicographic likelihood" ("leximax" likelihood for short). The leximax likelihood relation can be generated from the well-ordered partition $\{S_1, ..., S_m\}$ of S induced by π, as follows (Dubois et al, 1998):

$$A >_{\Pi Lex} B \text{ iff there is an } S_k \text{ such that } |B \cap (S_1 \cup...\cup S_k)| < |A \cap (S_1 \cup...\cup S_k)|$$

$$\text{and } |A \cap (S_1 \cup...\cup S_j)| = |B \cap (S_1 \cup...\cup S_j)| \text{ for all } j < k.$$

The relation $\geq_{\Pi Lex}$ is a complete preordering of events whose strict part refines the possibilistic likelihood $>_{\Pi L}$. In fact, the leximax likelihood relation coincides with the possibilistic likelihood for linear possibility distributions. For a uniform possibility distribution the leximax likelihood relation coincides with the comparative probability relation that is induced by a uniform probability (the cardinality-based definition gives $A >_{\Pi Lex} B$ iff $|B| < |A|$). This is not surprising in view of the fact that the leximax likelihood relation is really a comparative probability relation in the usual sense.

Unsurprisingly, in view of the above, any leximax likelihood relation on a finite set can be represented by a special kind of probability measure. Indeed, consider the well-ordered partition $\{S_1, ..., S_m\}$ of S. Let $r_m = 1$ and $r_i = |S_{i+1}| \cdot r_{1+1} + 1$, for $i < m$. Define a probability distribution p such that p(s) is proportional to r_i if $s \in S_i$. Then, it can be proved that:

$$B \geq_{\Pi Lex} C \;\square\; P(B) \geq P(C). \tag{26}$$

These probabilities are uniformly distributed inside the sets S_i and such that if $s \in S_i$ then $p(s) > P(S_{i+1}\cup...\cup S_m)$, i.e. p(s) is greater than the sum of the probabilities of all less probable states. Such probabilities generalize the so-called "big-stepped" probabilities (when the S_i are singletons, see Snow (1994, 1999) and also studied in Benferhat et al. (1999)). They must be used when refining possibilistic criteria by means of discrete expected utilities after Fargier and Sabbadin (2003).

6. Conclusion

This paper has given an account of qualitative decision rules under uncertainty. They can be useful for solving discrete decision problems involving finite state spaces and where it is not natural or very difficult to quantify utility functions or probabilities. For instance, there is no time granted to do it because a quick advice must be given (recommender systems). Or the problem takes place in a dynamic environment with a large state space, a non-quantifiable goal to be pursued, and partial information on the current state (autonomous vehicles); see Sabbadin, 2000. Or yet a very high level description of a decision problem is available, where states and consequences are roughly described (strategic decisions). The possibilistic criteria have also been used in scheduling, when the goal is to produce robust schedules for jobs, ensuring limited balanced violations of due-dates (Dubois et al. 1995). They are compatible with dynamic programming algorithms for multiple-stage decision problems (Fargier et al., 1999).

Two kinds of decision rules have been found. Some are consistent with the Pareto ordering and satisfy the sure thing principle, but leave room to incomparable decisions, and overfocus on most plausible states. The other ones do rank decisions, but lack discrimination. It seems that there is a conflict between fine-grained discrimination and the requirement of a total ordering in the qualitative setting. Future works should strive towards exploiting the complementarities of prioritized Pareto-efficient decision methods as the likely dominance rule, and of the pessimistic decision rules related to the maximin criterion. Putting these two requirements together seems to lead us back to a very special case of expected utility, as very recent results seem to indicate. Besides, the likely dominance rule compares two acts independently of other ones. One may imagine more expressive decision rules comparing two acts on the basis of a third one, thus giving up a property of irrelevance of other alternatives implicit in the ordinal invariance axiom. Lastly, our results also apply, with minor adaptation, to multicriteria decision-making, where the various objectives play the role of the states and the likelihood relation is used to compare the relative importance of groups of objectives (Dubois et al, 2001).

References

Arrow K. (1951). *Social Choice and Individual Values*. New York, N.Y.: Wiley.

Arrow, K. Hurwicz L. (1972). An optimality criterion for decision-making under ignorance. in: C.F. Carter, J.L. Ford, eds., *Uncertainty and Expectations in Economics*. Oxford, UK: Basil Blackwell & Mott Ltd.

Bacchus F. and Grove A. (1996). Utility independence in a qualitative decision theory. In *Proc. of the 5rd Inter. Conf. on Principles of Knowledge Representation and Reasoning (KR'96)*, Cambridge, Mass., 542-552.

Benferhat S., Dubois D., Prade H. (1999) Possibilistic and standard probabilistic semantics of conditional knowledge bases. *J. Logic & Computation*, 9, 873-895.

Boutilier C. (1994). Towards a logic for qualitative decision theory. In *Proc. of the 4rd Inter. Conf. on Principles of Knowledge Representation and Reasoning (KR'94)*, Bonn, Germany, May. 24-27, 75-86.

Bouyssou D., Marchant T., Pirlot M., Perny P., Tsoukias A. and Vincke P. (2000). *Evaluation Models: a Critical Perspective*. Kluwer Acad. Pub. Boston.

Brafman R.I., Tennenholtz M. (1997). Modeling agents as qualitative decision makers. *Artificial Intelligence*, 94, 217-268.

Brafman R.I., Tennenholtz M. (2000). On the Axiomatization of Qualitative Decision Criteria, *J. ACM*, 47, 452-482

Buckley J. J. (1988) Possibility and necessity in optimization, *Fuzzy Sets and Systems*, 25, 1-13.

Cayrol M., Farreny H. (1982). Fuzzy pattern matching. *Kybernetes*, 11, 103-116.

Cohen M. and Jaffray J.-Y. (1980) Rational behavior under complete ignorance. *Econometrica*, 48, 1280-1299.

Doyle J., Thomason R. (1999). Background to qualitative decision theory. *The AI Magazine*, 20 (2), Summer 1999, 55-68

Doyle J., Wellman M.P. (1991). Impediments to universal preference-based default theories. *Artificial Intelligence*, 49, 97- 128.

Dubois D. (1986). Belief structures, possibility theory and decomposable confidence measures on finite sets. *Computers and Artificial Intelligence* (Bratislava), 5(5), 403-416.

Dubois D. (1987) Linear programming with fuzzy data, In: *The Analysis of Fuzzy Information — Vol. 3: Applications in Engineering and Science* (J.C. Bezdek, ed.), CRC Press, Boca Raton, Fl., 241-263

Dubois D., Fargier H., and Prade H. Fuzzy constraints in job-shop scheduling. *J. of Intelligent Manufacturing*, 64:215-234, 1995.

Dubois D., Fargier H., and Prade H. (1996). Refinements of the maximin approach to decision-making in fuzzy environment. *Fuzzy Sets and Systems*, 81, 103-122.

Dubois D., Fargier H., and Prade H. (1997). Decision making under ordinal preferences and uncertainty. *Proc. of the 13th Conf. on Uncertainty in Artificial Intelligence* (D. Geiger, P.P. Shenoy, eds.), Providence, RI, Morgan & Kaufmann, San Francisco, CA, 157-164.

Dubois D., Fargier H., and Perny P. (2001). Towards a qualitative multicriteria decision theory. *Proceedings of Eurofuse Workshop on Preference Modelling and Applications*, Granada, Spain, 121-129, 25-27 avril 2001. To appear in Int. J. Intelligent Systems, 2003.

Dubois D., Fargier H., Perny P. and Prade H. (2002a). Qualitative Decision Theory: from Savage's Axioms to Non-Monotonic Reasoning. *Journal of the ACM*, 49, 455-495.

Dubois D., Fargier H., and Perny P. (2002b). On the limitations of ordinal approaches to decision-making. *Proc. of the 8th International Conference, Principles of Knowledge Representation and Reasoning* (KR2002), Toulouse, France. Morgan Kaufmann Publishers, San Francisco, California, 133-144.

Dubois D., Fargier H., and Perny P. (2003). Qualitative models for decision under uncertainty: an axiomatic approach. *Artificial Intell.*, to appear.

Dubois D., Fortemps P. (1999). Computing improved optimal solutions to max-min flexible constraint satisfaction problems. *European Journal of Operational Research,* 118, p. 95--126.

Dubois D., Grabisch M., Modave F., Prade H. (2000) Relating decision under uncertainty and multicriteria decision making models. *Int. J. Intelligent Systems*, 15 n°10, 967--979.

Dubois D., Marichal J.L., Prade H., Roubens M., Sabbadin R. (2001) The use of the discrete Sugeno integral in decision-making: a survey. *Int. J. Uncertainty, Fuzziness and Knowledge-based Systems*, 9, 539-561.

Dubois D., Prade H. (1988) *Possibility Theory — An Approach to the Computerized Processing of Uncertainty*. Plenum Press, New York

Dubois D., Prade H. (1995a) Numerical representation of acceptance. *Proc. of the 11th Conf. on Uncertainty in Articicial Intelligence*, Montréal, August, 149-156.

Dubois D., Prade H.(1995b) Possibility theory as a basis for qualitative decision theory. *Proc. of the Inter. Joint Conf. on Artificial Intelligence (IJCAI'95)*, Montréal, August, 1924-1930.

Dubois D., Prade H. (1998). Possibility theory: qualitative and quantitative aspects. P. Smets, editor, *Handbook on Defeasible Reasoning and Uncertainty Management Systems — Volume 1: Quantified Representation of Uncertainty and Imprecision*. Kluwer Academic Publ., Dordrecht, The Netherlands, 169-226

Dubois D., Prade H., and Sabbadin R. (1998). Qualitative decision theory with Sugeno integrals.*Proceedings of 14th Conference on Information Processing and Management of Uncertainty in Artificial Intelligence* (UAI'98), Madison, WI, USA. Morgan Kaufmann, San Francisco, CA, p. 121--128.

Dubois D., Prade H., and Sabbadin R. (2001). Decision-theoretic foundations of possibility theory. *European Journal of Operational Research*, 128, 459-478.

Dubois D., Prade H., Testemale C. (1988). Weighted fuzzy pattern matching. *Fuzzy Sets and Systems*, 28, 313-331.

Fargier H., Lang J. and Schiex T. (1993) Selecting preferred solutions in Fuzzy Constraint Satisfaction Problems, *Proc. of the 1st Europ. Conf. on Fuzzy Information Technologies (EUFIT'93)*, Aachen, Germany, 1128-1134.

Fargier H., Perny P. (1999). Qualitative models for decision under uncertainty without the commensurability assumption. *Proc. of the 15th Conf. on Uncertainty in Artificial Intelligence* (K. Laskey, H. Prade, eds.), Providence, RI, Morgan & Kaufmann, San Francisco, CA, 157-164.

Fargier H., Lang J., Sabbadin R. (1998). Towards qualitative approaches to multi-stage decision making. *International Journal of Approximate Reasoning*, 19, 441—471.

Fargier H., Sabbadin R., (2000) Can qualitative utility criteria obey the surething principle? *Proceedings IPMU2000*, Madrid, 821-826.

Fargier H. Sabbadin R. (2003) Qualitative decision under uncertainty: back to expected utility, *Proc. IJCAI'03*, Acapulco, Mexico.

Fishburn P. (1975). Axioms for lexicographic preferences. *Review of Economical Studies*, 42, 415-419

Fishburn P. (1986). The axioms of subjective probabilities. *Statistical Science* 1, 335-358.

Friedman N., Halpern J. (1996). Plausibility measures and default reasoning. *Proc of the 13th National Conf. on Artificial Intelligence (AAAI'96)*, Portland, 1297-1304.

Giang P., Shenoy P. (2000). A qualitative utility theory for Spohn's theory of epistemic beliefs. *Proc. of the 16th Conf. on Uncertainty in Artificial Intelligence*, 220-229.

Giang P., Shenoy P. (2001). A comparison of axiomatic approaches to qualitative decision-making using possibility theory. *Proc. 17th Int. Conf. on Uncertainty in Artificial Intelligence*, 162-170.

Grabisch M., Murofushi T., Sugeno M., Eds. (1999) *Fuzzy Measures and Integrals* Physica-Verlag, Heidelberg, Germany.

Grabisch M., De Baets B., and Fodor J. (2002) On symmetric pseudo-additions and pseudo-multiplications: is it possible to build rings on [-1, +1]? In *Proc. 9th Int Conf. on Information Processing and Management of Uncertainty in Knowledge based Systems (IPMU2002)*, Annecy, France, pp 1349-1355.

Grant S., Kajii A., Polak B.(2000) Decomposable Choice under Uncertainty, *J. Economic Theory*. Vol. 92, No. 2, pp. 169-197.

Jaffray J.-Y. (1989) Linear utility theory for belief functions. *Operations Research Letters*, 8, 107-112.

Inuiguichi M., Ichihashi H., and Tanaka.H. (1989). Possibilistic linear programming with measurable multiattribute value functions. *ORSA J. on Computing*, 1, 146-158.

Kraus K., Lehmann D., Magidor M. (1990). Nonmonotonic reasoning, preferential models and cumulative logics. *Artificial Intelligence*, 44, 167-207.

Lang J. (1996). Conditional desires and utilities: an alternative logical approach to qualitative decision theory. *Proc. of the 12th European Conf. on Artificial Intelligence (ECAI'96)*, Budapest, 318-322.

Lehmann D. (1996). Generalized qualitative probability: Savage revisited. *Proc. of the 12th Conf. on Uncertainty in Artificial Intelligence*, Portland, August, Morgan & Kaufman, San Mateo, CA, 381-388.

Lehmann D. (2001). Expected Qualitative Utility Maximization, *J. Games and Economic Behavior*. 35, 54-79

Lewis D. (1973). *Counterfactuals*. Basil Blackwell, London.

Moulin H. (1988). *Axioms of Cooperative Decision Making*. Cambridge University Press, Cambridge, MA.

Roubens M., Vincke P.(1985) *Preference Modelling*. Lecture Notes in Economics and Mathematical Systems, Vol. 250, Springer Verlag, Berlin.

Sabbadin R. (2000), Empirical comparison of probabilistic and possibilistic Markov decision processes algorithms. *Proc. 14th Europ. Conf. on Artificial Intelligence (ECAI'00)*, Berlin, Germany, 586-590.

Savage L.J. (1972). *The Foundations of Statistics*. Dover, New York.

Schmeidler D. (1989) Subjective probability and expected utility without additivity, *Econometrica*, 57, 571-587.

Sen A.K. (1986). Social choice theory. In K. Arrow, M.D. Intrilligator, Eds., *Handbook of Mathematical Economics*, Chap. 22, Elsevier, Amsterdam, 1173-1181.

Shackle G.L.S. (1961) *Decision Order & Time In Human Affairs*, Cambridge University Press, Cambridge, U.K .(2nd edition, 1969).

Snow P. (1999) Diverse confidence levels in a probabilistic semantics for conditional logics. *Artificial Intelligence* 113, 269-279.

Tan S.W., Pearl J. (1994). Qualitative decision theory. *Proc. 11th National Conf. on Artificial Intelligence (AAAI-94)*, Seattle, WA, pp. 70-75.

Thomason R. (2000), Desires and defaults: a framework for planning with inferred goals. *Proc. of the Inter. Conf. on Principles of Knowledge Representation and Reasoning (KR'00)*, Breckenridge, Col.., Morgan & Kaufmann, San Francisco, 702-713.

Von Neumann J. and Morgenstern O. (1944): *Theory of Games and Economic Behaviour* (Princeton Univ. Press, Princeton, NJ).

Vincke P. *Multicriteria Decision-Aid*, J. Wiley & Sons, New York, 1992

Wald A. (1950), *Statistical Decision Functions*. J. Wiley & Sons, New York.

Whalen T. (1984). Decision making under uncertainty with various assumptions about available information. *IEEE Trans. on Systems, Man and Cybernetics*, 14:888-900.

Yager.R.R. (1979). Possibilistic decision making. *IEEE Trans. on Systems, Man and Cybernetics*, 9:388-392,.

L.A. Zadeh L.A. (1978). Fuzzy sets as a basis for a theory of possibility. *Fuzzy Sets and Systems*, 1, 3-28.

Sequential Decision Making in Heuristic Search

Eyke Hüllermeier

Department of Mathematics and Informatics
University of Marburg
Germany

Abstract. The order in which nodes are explored in a (depth-first) iterative deepening search strategy is principally determined by the condition under which a path of the search tree is cut off in each search phase. A corresponding criterion, which has a strong influence on the performance of the overall (heuristic) search procedure, is generally realized in the form of an upper cost bound. In this paper, we develop an effective and computationally efficient termination criterion based on statistical methods of change detection. The criterion is local in the sense that it depends on properties of a path itself, rather than on the comparison with other paths. Loosely speaking, the idea is to take a systematic change in the (heuristic) evaluation of nodes along a search path as an indication of suboptimality. An expected utility criterion which also takes the consequence of the suboptimal search decision on the solution quality into account is proposed as a generalization of this idea.

1 Introduction

Heuristic search strategies (for single-agent path-finding problems) explore more promising paths of a search tree (or, more generally, a search graph) before less promising ones by evaluating search states, thereby putting the successors of an inner node in some order according to their desirability. A frequently used type of (numeric) evaluation function is of the form

$$f(\eta) = g(\eta) + h(\eta),$$

where $g(\eta)$ denotes the cost of the path from the root (initial state) to the node η and $h(\eta)$ is an estimation of the cost, $h^*(\eta)$, of the shortest (= lowest-cost) path from η to some goal state. The value $h(\eta)$ is generally derived from some "features" of the search state associated with η. Considerable attention has been payed to so-called *admissible* heuristics which underestimate the true cost function h^*, i.e., which satisfy $h(\eta) \leq h^*(\eta)$ for all nodes η. Such heuristics are of interest since they allow for realizing search algorithms which are guaranteed to return an optimal (lowest-cost) solution.

One possibility of deriving admissible heuristics in a systematic way is to use simplified (less constrained) models of the original problem [15]. Still, the discovery of good admissible heuristics remains a difficult task, let alone the automatization of this process [17]. Besides, the computation of relaxation-based heuristic functions might be expensive since it calls for solving new, even though simplified, problems. In this

paper, we assume heuristic information which does not provide lower bounds but estimations of cost values in a proper (statistical) sense. The (data-oriented) acquisition of
this type of heuristic function seems to be less difficult than the (knowledge-oriented)
invention of admissible heuristics. For instance, statistical techniques can be used for
approximating the cost function h^* from a set of training data. Of course, when basing
a (best-first) search strategy on inadmissible heuristics, the (first) solution found might
not be optimal. Nevertheless, corresponding evaluation functions are often more accu-
rate, and using them in lieu of admissible ones may yield a considerable improvement
in (average) time complexity at the cost of an acceptable (perhaps provably bounded
[6]) deterioration of solution quality.

It is well-known that (depth-first) iterative deepening (ID) search can overcome the
main limitations of the basic breadth-first (BFS) and depth-first (DFS) search strategies, namely the exponential space complexity of the former and the non-optimality
and non-completeness of the latter [12]. The criterion used by ID for cutting off search
paths is generally given in the form of a global depth-limit or cost-limit. In this paper,
we propose alternative termination criteria based on statistical methods of *change detection*. Loosely speaking, the idea is not to use the evaluation of individual (frontier)
nodes itself but the *change* in the node evaluation along a search path in order to detect
(and cut off) hardly promising paths.

The paper is organized as follows: In Section 2, we present a motivating example
and potential application of our search method, namely knowledge-based configuration.
In Section 3, we explain the problem of change detection in its general form. Then,
some basic algorithms for approaching this problem are briefly reviewed in Section 4.
The application of change detection in the context of heuristic search is discussed in
Section 5. Finally, a generalized approach to search termination based on an expected
utility criterion is proposed in Section 6. The paper concludes with a few remarks.

2   Motivation: Configuration as Heuristic Search

As a motivating example let us consider *resource-based configuration* (RBC), a special approach to knowledge-based configuration [7]. It proceeds from the idea that a
(technical) system is assembled from a set of primitive *components*. A resource-based
description of components is a special type of property-based description in which each
component (e.g. a lamp) is characterized by some set of *resources* or *functionalities*
it provides (e.g. light) and some other set of resources it demands (e.g. electric current). The relation between components is modeled in an abstract way as the exchange
of resources (cf. Fig. 1). A configuration problem consists of minimizing the price of
a configuration while satisfying an external demand of functionalities. In its simplest
form it is specified as a triple $\langle A, d, c \rangle$, where A is a set of components and d is an
external demand of functionalities. Each component is characterized by some integer
vector $a = (a_1, \dots, a_m)$ with the intended meaning that it offers f_i, i.e. the ith functionality, a_i times if $a_i > 0$ $(1 \leq i \leq m)$. Likewise, the component demands this
functionality a_i times if $a_i < 0$. The set of components can be written compactly in
the form of an $m \times n$ integer matrix which we also refer to as A. The jth column a^j
of A corresponds to the vector characterizing the jth component. The external demand

is also specified as a vector $d \geq 0$, and the meaning of its entries d_i is the same as for the components except for the sign. The (integer) vector $c = (c_1, \ldots, c_n)$ defines the prices of the components. Using the jth component k times within a configuration causes costs of $k \cdot c_j > 0$. A configuration, i.e. the composition of a set of components, is written as a vector $x = (x_1, \ldots, x_n)$ with x_j being the number of occurrences of the jth component. A configuration x is *feasible* if the net result of the corresponding composition and the external demand d are "balanced." Using the notation introduced above, this condition can be written

$$\sum_{j=1}^{n} x_j \, a^j \geq d.$$

A feasible configuration (solution) x^* is called optimal if it causes minimal cost. In its basic form a resource-based configuration problem is obviously equivalent to an integer linear program

$$A \times x \geq d, \quad c \times x \rightarrow min.$$

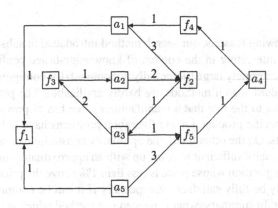

Fig. 1. Dependency graph of an RBC problem indicating the offer (directed edge from component a_i to functionality f_j) and demand (directed edge in the reverse direction) of functionalities.

Due to this equivalence, one could think of using standard methods from operations research for solving it. However, this equivalence is already lost under slight (but practically relevant) generalizations of the basic model [3]. Realizing a heuristic search in the *configuration space*, i.e. the set of possible configurations (identified by integer-valued vectors), seems to be a reasonable alternative which is more robust against extensions of the model. In the corresponding search tree, each node η is associated with a partial configuration $x(\eta)$, and successor nodes are obtained by adding components to that configuration (cf. Fig. 2).

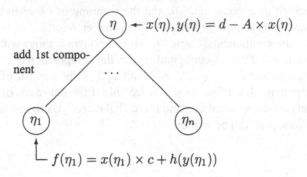

Fig. 2. Configuration as heuristic search: Each node η of the search tree is associated with an intermediate configuration $x(\eta)$. The kth successor of η is obtained by adding the kth component to $x(\eta)$. It is evaluated according to the estimated cost of the optimal solution in the corresponding subtree, namely the sum of the cost of $x(\eta_k)$ and the estimated cost of the remaining demand $y(\eta_k)$. These evaluations determine the order in which the successors $\eta_1, \ldots, \eta_n$ are expanded.

Due to the following reasons, our search method introduced in subsequent sections appears especially interesting in the context of knowledge-based configuration. First, the search space is extremely large (potentially unbounded) for configuration problems. Consequently, standard search methods are hardly applicable. The problem becomes especially severe due to the fact that a manufacturer often has to provide an "on-line" offer for a client-specific product, i.e. configuration problems have to be solved under real-time constraints. On the other hand, the optimality constraint can usually be weakened. That is, it is usually sufficient to come up with an approximately optimal solution. For instance, a configuration whose price is less than 1% above the price of the optimal solution will usually be fully satisfactory, especially if it can be computed 1000 times faster than the latter. In summary, what is needed is a method which is able to trade off efficiency against accuracy, and which provides a good solution in reasonable running time. Exactly this principle is supported by our approach.

Finally, it turned out that the type of heuristic evaluation function assumed by our search method is indeed much simpler to obtain for RBC problems than admissible heuristics. In fact, by applying techniques from statistical regression to a training set of optimally solved configuration problems[1] we were able to derive very good estimations h of the cost function h^* [9]. For each node η of the search tree, $h(\eta)$ is an estimation of the cost of the demand $d - A \times x(\eta)$ that remains of the original demand d (associated with the root of the tree). More specifically, $h(\eta)$ is an unbiased estimation in a statistical sense and follows an approximate normal distribution.

---

[1] Of course, optimal solutions have been derived "off-line".

The combination of this statistical approach for estimating cost functions and the search strategy introduced below has lead to a very effective method for solving RBC problems. Due to reasons of space, and since the focus of the paper is on methodological aspects of heuristic search, we refrain from discussing further technical details of this application and rather refer to the references [9, 11].

3 The Problem of Change Detection

The problem of change detection is understood as the detection of *abrupt changes* in some characteristic properties of a system. Generally, such characteristics are not observed directly (e.g., through corresponding sensors) but can only be inferred indirectly from the measurements available. The problem of change detection arises, e.g., in pattern recognition (segmentation of signals), quality control, monitoring in biomedicine, or in connection with the detection of faults in technological processes [1].

Many problems of this type can be stated as one of detecting a change in the *parameters* of a (static or dynamic) stochastic model. One possibility of approaching change detection is hence from the viewpoint of mathematical statistics. Consider, for instance, a (discrete) sequence $(Y_t)_{1 \leq t \leq T}$ of random variables, where Y_t is characterized by the (parametrized) conditional probability (density) $\phi_\theta(\cdot \mid y_{t-1}, \ldots, y_1)$. We suppose that a change is reflected by the change of the parameter vector θ. Thus, assuming that the random variables are independent of each other and that at most one change has occured, we have

$$\phi(y_t) = \begin{cases} \phi_{\theta_0}(y_t) & \text{if } 1 \leq t < t_0 \\ \phi_{\theta_1}(y_t) & \text{if } t_0 \leq t \leq T \end{cases},$$

where t_0 denotes the change time. The problem of change detection can comprise different tasks: (a) Detecting a change as soon as possible, i.e., deciding at each point of time whether the sequence observed so far contains a change or not. (b) Estimating the change time t_0, if any. (c) Estimating the (possibly unknown) parameters θ_0 and θ_1. These problems can be formalized within the framework of mathematical statistics. The first task, for instance, can be considered as a problem of hypotheses testing: The hypothesis H_0 that $\theta(t) = \theta_0$ for $1 \leq t \leq T$, where $\theta(t)$ denotes the parameter at time t, is tested against the alternative, H_1, that

$$\exists 1 < t_0 \leq T : \theta(1) = \ldots = \theta(t_0 - 1) = \theta_0$$
$$\wedge \ \theta(t_0) = \ldots = \theta(T) = \theta_1.$$

(This test assumes θ_0 and θ_1 to be known.) Likewise, the second task can be approached as a problem of estimating a (discrete) parameter, namely t_0. Note that the second and third task can be considered as auxiliary problems, once the first question has been answered in favor of H_1.

Different performance measures for assessing change detection procedures exist. Even though the importance of a criterion depends on the respective application, the following criteria are relevant in any case: The *expected delay* of a detection is the expected number of time points between t_0 and the time when the change is detected.

The *probability of a false detection* is the probability of deciding on H_1 even though H_0 is true. These criteria are irreconcilable, of course: Reducing the probability of a false detection will generally increase the expected delay and vice versa.

4 Algorithms for Change Detection

In this section, we review one of the basic procedures for change detection, the so-called cumulative sum (CUSUM) algorithm. It is based upon the repeated use of the sequential probability ratio test (SPRT). The CUSUM algorithm was first proposed in [14].

4.1 The Sequential Probability Ratio Test

The SPRT, investigated in depth by WALD [18], is one of the most important test procedures in sequential statistical analysis, mainly due to its optimality properties [19] and its computational efficiency. It is a test for deciding between two simple statistical hypotheses

$$H_0 : \theta = \theta_0, \quad H_1 : \theta = \theta_1 \tag{1}$$

about some (constant) parameter θ. Consider a sequence $(Y_t)_{t \geq 1}$ of independent and identically distributed random variables, and let ϕ_{θ_0} and ϕ_{θ_1} denote the probability (density) function of Y_t valid under H_0 and H_1, respectively. Moreover, suppose that this sequence has been observed up to time T. A value of the log-likelihood ratio (the logarithm of the likelihood ratio)

$$\mathcal{L}_T = \sum_{t=1}^{T} \ln(\phi_{\theta_1}(y_t)) - \ln(\phi_{\theta_0}(y_t)) \tag{2}$$

larger (smaller) than 0 can then be interpreted as an indication of the validity of H_1 (H_0). An intuitive decision rule is hence to accept H_0, if $\mathcal{L}_T$ falls below some threshold $\alpha < 0$, and to reject H_0 (i.e., to decide on H_1), if $\mathcal{L}_T$ exceeds a threshold $\beta > 0$. The SPRT realizes this idea by combining a stopping rule and a terminal decision rule:

```
while α < L_T < β
    continue sampling (T = T + 1)
if L_T ≤ α then accept H_0
if β ≤ L_T then reject H_0
```

Observe that the log-likelihood ratio can be written recursively as $\mathcal{L}_T = \mathcal{L}_{T-1} + z_T$, where $z_T = \ln(\phi_{\theta_1}(y_T)) - \ln(\phi_{\theta_0}(y_T))$. Loosely speaking, the SPRT takes observations as long as evidence in favor of either H_0 or H_1 is not convincing enough. The thresholds α and β are directly related to the probabilities of two types of errors, namely of falsely rejecting H_0 and of deciding in favor of H_0 even though H_1 is true. Indeed, α and β can be used for controlling these errors. Of course, a tradeoff between a small probability of an error and the quickness of a decision, i.e., the length of the observed sequence, has to be achieved: The smaller (larger) α (β) is, the smaller is the probability of an error. At the same time, however, the terminal decision will be delayed, i.e., the *exit time*, T_0, will be larger.

4.2 The CUSUM Algorithm

The SPRT assumes that the complete data collected during the test is generated by one and the same model. Thus, it has to be modified in the context of change detection, where the data-generating process is assumed to change in-between.

Again, we are interested in testing the hypotheses (1) repeatedly. Now, however, H_0 is definitely true at the beginning, and the sampling is stopped only if a change seems to have occured, i.e., if the test statistic exceeds an upper bound. In order to avoid a detection delay due to the fact that (2) decreases as long as H_0 holds true it is hence reasonable to restrict the accumulated evidence in favor of H_0. This can be achieved by letting α define a lower bound to the test statistic. When choosing $\alpha = 0$ [13], the CUSUM algorithm can be written in its basic form as follows:

$$
\begin{aligned}
&g_0 = 0, \quad T = 0 \\
&\texttt{repeat until } g_T \geq \beta \\
&\quad T = T + 1 \\
&\quad g_T = \max\{0, g_{T-1} + \ln(\phi_{\theta_1}(y_T)) - \ln(\phi_{\theta_0}(y_T))\}
\end{aligned}
$$

If the CUSUM algorithm has terminated at time T_0, the maximum likelihood (ML) estimation of the change time t_0 is given by

$$
\hat{t}_0 = \arg \max_{1 < t \leq T_0} \sum_{k=t}^{T_0} \ln(\phi_{\theta_1}(y_k)) - \ln(\phi_{\theta_0}(y_k)).
$$

Often, not only the change time t_0 is unknown but also the parameters θ_0 and θ_1. A standard statistical approach, then, is to use the corresponding ML estimations instead. This leads to

$$
g_T = \max_{1 < t_0 \leq T} \mathcal{L}_{t_0}(\hat{\theta}_{0,t_0}, \hat{\theta}_{1,t_0}) \tag{3}
$$

$$
= \max_{1 < t_0 \leq T} \sum_{t=t_0}^{T} \ln(\phi_{\hat{\theta}_{1,t_0}}(y_t)) - \ln(\phi_{\hat{\theta}_{0,t_0}}(y_t)),
$$

where $\hat{\theta}_{0,t_0}$ and $\hat{\theta}_{1,t_0}$ denote, respectively, the ML estimations of θ_0 and θ_1 based on the observations $y_1, \ldots, y_T$ and the assumption that t_0 is a change time. $\mathcal{L}_{t_0}(\theta_0, \theta_1)$ is the log-likelihood ratio associated with the hypotheses

$$
H_0 : \forall 1 \leq t \leq T : \theta(t) = \theta_0,
$$

$$
H_1 : \theta(t) = \begin{cases} \theta_0 & \text{if } 1 \leq t \leq t_0 - 1 \\ \theta_1 & \text{if } t_0 \leq t \leq T \end{cases}.
$$

Again, the stopping rule is defined by $g_T \geq \beta$, with β being a predefined threshold. That is, the exit time T_0 is defined as the smallest T such that $g_T \geq \beta$. The (ML) estimation of t_0 is then given by the change time $1 < \hat{t}_0 \leq T_0$ for which the maximum in (3) is attained. Observe that g_T, as defined in (3), can no longer be determined by means of a simple recursive formula.

5 Changes in Heuristic Search

In this section, we shall apply the idea of change detection to heuristic tree-search algorithms. We assume heuristic information to be available in the form of an estimation h of the cost function h^* which assigns to each node η the cost of the (cost-)optimal path from η to a goal state. A value $h(\eta)$ is considered as the realization of a random variable $H(\eta)$ [5]. We suppose the random variables associated with different search states to be independent of each other, a simplifying assumption commonly made in the probabilistic analysis of search trees [20]. For the sake of simplicity, we also assume $\mathsf{E}(H(\eta)) = h^*(\eta)$ and $(H(\eta) - h^*(\eta)) \sim \Phi$ for all nodes η, where E denotes the expected value operator.[2] Statistically speaking, $h(\eta)$ is an unbiased estimation of $h^*(\eta)$. Φ is a distribution with associated probability (density) function ϕ, i.e., $\phi(x)$ is the probability (density) that $H(\eta) - h^*(\eta) = x$.

The overall cost associated with a node η is given by $f(\eta) = g(\eta) + h(\eta)$, where $g(\eta)$ denotes the cost of the path from the initial state to η. The evaluation function f is used for guiding the search process. According to our assumptions above, $f(\eta)$ can be interpreted as the realization of a random variable $F(\eta)$ such that $(F(\eta) - g(\eta) - h^*(\eta)) \sim \Phi$.

5.1 Path Profiles

Consider a search path ρ, i.e., a sequence $(\eta_1, \ldots, \eta_K)$ of nodes with η_1 being the root of the search tree and η_K the (current) frontier node. We associate sequences

$$f(\rho) = (f(\eta_1), \ldots, f(\eta_K)),$$
$$f^*(\rho) = (f^*(\eta_1), \ldots, f^*(\eta_K))$$

with ρ, where $f^*(\eta_k) = g(\eta_k) + h^*(\eta_k)$. $f(\rho)$ is called the *path profile* of ρ. The sequence $f^*(\rho)$ is obviously non-decreasing, i.e.,

$$f^*(\eta_1) \leq f^*(\eta_2) \leq \ldots \leq f^*(\eta_K). \tag{4}$$

Suppose that we have not made a suboptimal search decision (node generation) so far, i.e., a cost-optimal solution can still be reached from η_K. Then, all inequalities in (4) are in fact equalities. That is, the random variables $F(\eta_k)$ have the same expectation and, hence, the same distribution. Now, suppose all but the kth decision to be optimal. Then,

$$\mu_0 = f^*(\eta_1) = f^*(\eta_2) = \ldots = f^*(\eta_k) \tag{5}$$
$$< f^*(\eta_{k+1}) = f^*(\eta_{k+2}) = \ldots = f^*(\eta_K) = \mu_1.$$

---

[2] This assumption can easily be relaxed.

That is, the expected value of the random variables $F(\eta_1), \ldots, F(\eta_k)$ is μ_0, whereas the expectation of $F(\eta_{k+1}), \ldots, F(\eta_K)$ is μ_1. More generally, we have the case that some (but at least one) of the inequalities in (4) are strict if not all decisions have been optimal. Subsequently, we assume that at most one suboptimal decision has been made. [3]

5.2 Change Detection and Search Termination

Iterative deepening search is the method of choice for many applications since it combines the merits of both, BFS (completeness, or even optimality) and DFS (linear space complexity). Taking a repeated (partially informed) depth-first search as a point of departure, the ID algorithm is principally determined by the conditions under which search along a path is broken off. The common approach is to use a depth-limit or, more generally, a cost-limit, for each search phase. For instance, the iterative deepening version of the A* algorithm, IDA*, continues search along a path as long as the frontier node η satisfies $f(\eta) \leq c$, with c being the f-cost-limit valid for the respective search phase.

When being interested in finding good solutions, or even optimal ones, one should obviously avoid the exploration of suboptimal search paths. Indeed, this idea is supported by the gradually increased cost-limit in IDA*. When being also interested in minimizing computational effort, it is reasonable, not only to *terminate* search along less promising paths, but also to *continue* search along a path as long as it appears promising, thereby leaving other (perhaps also promising) paths out of account. This idea is actually not supported by the use of an f-cost-limit which brings about a permanent comparison between a (large) set of simultaneously explored paths. [4] In fact, "following a search path as long as it might lead to a (near-)optimal solution" principally requires a *local* termination criterion which makes do with (heuristic) information from nodes of an *individual* path. Here, we shall use statistical methods of change detection in order to realize corresponding termination criteria. As will be seen, these methods allow for estimating the quality of a solution eventually found when following a path without referring to other search paths, or other solutions already encountered.

Consider a path $\rho = (\eta_1, \ldots, \eta_K)$ with path profile $f(\rho) = (f(\eta_1), \ldots, f(\eta_K))$. The parameter of interest, i.e., the change of which has to be detected, is the expected value $E(F(\eta_k))$. Depending on the information available about μ_0 and μ_1, the expected values before and after a change, different algorithms can be applied. Generally, it must be assumed that neither μ_0 nor μ_1 is known. Consequently, the basic CUSUM algorithm cannot be used directly. Rather, we have to proceed from (3). Suppose $1 < k_0 \leq K$ to be a changepoint, i.e., the generation of η_{k_0} was not optimal. The ML estimations $\hat{\mu}_{0,k_0}$ and $\hat{\mu}_{1,k_0}$ of μ_0 and μ_1 are then given by the mean of the values $f(\eta_1), \ldots, f(\eta_{k_0-1})$ and $f(\eta_{k_0}), \ldots, f(\eta_K)$, respectively. The test statistic (3)

---

[3] Experimental studies have shown that a generalization of this assumption hardly improves the quality of search decisions.

[4] Still, the depth-first component in A* can be strengthened by using a weighted evaluation function [16].

thus becomes

$$\mathcal{L}(\rho) = \max_{1 < k_0 \leq K} \mathcal{L}_{k_0}(\rho)$$

$$= \max_{1 < k_0 \leq K} \sum_{k=k_0}^{K} \ln(\phi_{\hat{\mu}_{1,k_0}}(f(\eta_k))) - \ln(\phi_{\hat{\mu}_{0,k_0}}(f(\eta_k))),$$

where ϕ_μ denotes the probability (density) function of the random variable $X + \mu$ and $X \sim \Phi$. Observe that we should actually use the *constrained* ML estimations of μ_0 and μ_1, taking into account that $\mu_0 \leq \mu_1$. In fact, the case where $\hat{\mu}_{0,k_0} > \hat{\mu}_{1,k_0}$ indicates an (actually impossible) decrease of the expectation and can hence be ignored. The termination criterion should thus be defined as

$$\mathcal{L}'(\rho) = \max_{1 < k_0 \leq K} \mathcal{L}'_{k_0}(\rho) \geq \beta, \tag{6}$$

where $\mathcal{L}'_{k_0}(\rho) = \mathcal{L}_{k_0}(\rho)$ if $\hat{\mu}_{0,k_0} \leq \hat{\mu}_{1,k_0}$ and 0 otherwise, and $\beta > 0$ is a predefined threshold. $\mathcal{L}'(\rho)$ can be computed in time $O(K)$ by the algorithm in Figure 3.

```
function change(f(η₁),...,f(η_K))
    m(1) = f(η₁)
    for k = 2 to K
      m(k) = m(k-1) + f(η_k)
    for k₀ = 2 to K
      μ̂₀,k₀ = m(k₀-1)/(k₀-1)
      μ̂₁,k₀ = (m(K) - m(k₀-1))/(K - k₀ + 1)
    L' = 0,  k̂₀ = 0,  s = 0
    for k₀ = K down to 2
      s = s + ln(φ_μ̂₁,k₀(f(η_k))) - ln(φ_μ̂₀,k₀(f(η_k)))
      if  μ̂₀,k₀ ≤ μ̂₁,k₀  and  L' < s  then
        L' = s,  k̂₀ = k₀
    return L', k̂₀, μ̂₀,k̂₀
```

Fig. 3. Pseudo-code for computing the termination criterion (6) and the ML estimations $\hat{k}_0$ and $\hat{\mu}_0$.

5.3 An Iterative Deepening Algorithm

When making use of the above termination criterion, a search algorithm cuts of the current path and starts a backtracking whenever (6) is satisfied (or a terminal node has been reached). As already mentioned before, the threshold β can be used for controlling two (conflicting) criteria: The *expected delay* of a change detection corresponds to the expected number of nodes explored along a path after a suboptimal search decision has been made. Of course, the smaller the threshold β is, the smaller is the detection delay. The probability of a *false detection* corresponds to the probability of cutting off an optimal search path. Keeping this probability small calls for a large threshold β.

Choosing $\beta = \infty$ leads to a pure depth-first search and, hence, excludes the probability of a false detection completely. In general, however, it may happen that an optimal search path is cut off and the corresponding search process terminates before having found a solution. Suppose, for instance, that a change has occured at $k_0 = 11$. The following table shows the (experimentally determined) probability p of terminating before $k = 11$ and the expected detection delay, d, for different thresholds β if Φ is the standard normal distribution and $\mu_1 = \mu_0 + 2$:

β	p	d	β	p	d
5	0.48	1.51	8	0.28	2.55
6	0.39	1.86	9	0.25	2.85
7	0.33	2.20	10	0.22	3.17

If the search process terminates without a solution, a new search phase with a larger threshold (which might be defined on the basis of the search process so far) has to be started. Based on an increasing sequence $(\beta_i)_{i \geq 0}$ of thresholds we thus obtain an iterative deepening algorithm, IDCD (Iterative Deepening based on Change Detection).

Of course, the first solution found by IDCD is not necessarily an optimal one. In fact, using the kind of heuristic information available does generally not allow for guaranteeing optimality, since the true cost value of a search node might not only be underestimated but also overestimated. It seems hence reasonable to let IDCD return the first solution found. Since the order in which solutions are encountered depends on the order in which (promising) search paths are explored, the successors of inner nodes should be arranged according to their f-evaluations. This way, a partially informed backtracking is realized in each search phase.

An interesting question concerns the completeness of IDCD. It can be shown that IDCD terminates and returns a solution with probability 1 under rather general conditions. This result can be proved by making use of a related termination property of the SPRT [18]. However, the proof becomes non-trivial due to the fact that one has to consider an infinite number of search paths.

It is worth mentioning that the change detection criterion suggests some kind of (probabilistic) backmarking strategy [4]: Instead of returning to the closest unexpanded ancestor, one may back up to the parent of the most likely changepoint $\hat{k}_0$ directly, thereby pruning (large) parts of the search tree. Of course, a corresponding algorithm will in general not guarantee completeness.

6 Utility-Based Search

The principle underlying IDCD is to continue a search path only if it might lead to an optimal solution. Now, suppose that we consider a *comprehensive* value of a computation [8] including the solution quality as well as the running time, and that we are

willing to gain efficiency at the cost of solution quality. In other words, we are interested in finding a "good" solution with reasonable computational effort, rather than
finding an optimal one regardless of the running time. In this section, we shall propose
a generalization of (6) suitable for supporting search under these assumptions.

Suppose the degree to which a user is satisfied with a solution, x, to depend on some
relation between the cost of x and the cost of an optimal solution, say, the difference
$\Delta(x)$ between these values. The preferences of the user can then be formalized by
means of a (non-increasing) utility function $U : [0, \infty) \to [0, 1]$, where $U(\Delta(x))$ is the
utility of a solution x.

Given the preferences of the user thus defined, a reasonable generalization of the
termination criterion (6) is to break off search along a path $\rho = (\eta_1, \ldots, \eta_K)$ if the *expected utility* of the best solution reachable from η_K, $x^*(\rho)$, falls below some threshold
u_0, i.e., if

$$\mathsf{E}(U(\rho)) = \int_0^\infty U(\Delta) \cdot \pi(\Delta \mid f(\rho)) \, d\Delta < u_0. \tag{7}$$

Here $\pi(\Delta \mid f(\rho))$ denotes the probability (density) that $\Delta(x^*(\rho)) = \Delta$, given the path
profile $f(\rho)$.

Suppose prior knowledge concerning Δ to be available in the form of a (prior)
probability (density) function π. The posterior $\pi(\cdot \mid f(\rho))$ is then specified by

$$\pi(\cdot \mid f(\rho)) \propto \lambda(f(\rho) \mid \cdot) \times \pi, \tag{8}$$

where $\lambda(f(\rho) \mid \Delta)$ is the probability of observing $f(\rho)$, given the deviation Δ. If we take
π as an uninformative prior,[5] then $\pi(\cdot \mid f(\rho))$ is proportional to the likelihood function
$\lambda(f(\rho) \mid \cdot)$, i.e., $\pi(\cdot \mid f(\rho)) \propto \lambda(f(\rho) \mid \cdot)$, where

$$\lambda(f(\rho) \mid \Delta) = \prod_{k=1}^{k_0-1} \phi_{\mu_0}(f(\eta_k)) \prod_{k=k_0}^{K} \phi_{\mu_0+\Delta}(f(\eta_k)). \tag{9}$$

The posterior $\pi(\cdot \mid f(\rho))$ cannot be derived from (8) directly if k_0 and μ_0 in (9) are
unknown. In Bayesian analysis, such parameters are called *nuisance parameters*. There
are different possibilities of deriving a posterior distribution for the parameters of interest in the presence of nuisance parameters [2]. One might consider, e.g., the marginal
distribution of the full posterior distribution which contains both, the parameters of interest as well as the nuisance parameters. This solution may become computationally
expensive, however. A further possibility is to replace the nuisance parameters by their
ML estimations. This approach seems reasonable in our situation since we can fall back
on the results obtained in connection with the detection of changes: Both, the estimation of the change point, $\hat{k}_0$, and of the corresponding expectation, $\hat{\mu}_0$, are computed by
the algorithm in Figure 3. A (utility-based) termination criterion can thus be realized as
follows:

– Derive $\hat{k}_0$ and $\hat{\mu}_0$ for the current search path ρ.

---

[5] Suppose $\pi(\cdot \mid f(\rho))$ to exist under this condition.

- Let $E(U(\rho)) = 1$ if $\hat{k}_0 = 0$. Otherwise, derive $E(U(\rho))$ according to (7), (8) and (9).
- Break off the search path if $E(U(\rho)) < u_0$, with $u_0 \leq 1$ being a predefined threshold.

This procedure reveals that the expected utility criterion can be seen as an extension of the change detection approach: Instead of only looking for a jump in the expected value, the decision whether to terminate search or not also depends on the size of this jump. The other way round, (6) corresponds to the special case of (7) in which the utility function is given by $U = 1_{\{1\}}$.

Observe that $\pi(\cdot \mid f(\rho))$ is not necessarily 0 on $(-\infty, 0)$ since (9) might be positive (even though small) also for $\Delta < 0$. Instead of incorporating the constraint $\Delta \geq 0$ by means of the usual conditioning (proportional allocation of probability mass) it seems reasonable to take the (actually impossible) event $\Delta < 0$ as evidence for $\Delta = 0$. Formally, this is equivalent to integrating over $(-\infty, \infty)$ in (7) with an extended utility function such that $U(\Delta) = U(0) = 1$ for all $\Delta < 0$.

By making use of (7) in conjunction with a sequence of utility thresholds $(u_i)_{i \geq 0}$ instead of (6) (and the sequence $(\beta_i)_{i \geq 0}$), we obtain an iterative deepening algorithm UBID (Utility-Based Iterative Deepening). Again, the search strategy is not only controlled by (7) but also by $(u_i)_{i \geq 0}$. A threshold $u_i = 0$, for instance, entails a pure depth-first search. The smaller the increments $\Delta u_i = u_i - u_{i-1}$ are defined, the more cautious search paths are explored, i.e., the more UBID resembles an (iterative deepening) best-first search.

Of course, since the expected utility criterion requires the computation of the integral (which is a sum in the case of a discrete distribution Φ) in (7) it is computationally more complex than the change detection criterion. Still, it can be realized very efficiently for several special cases. For instance, if Φ is a normal distribution[6] and U corresponds to an (extended) $\{0, 1\}$-valued utility function $1_{(-\infty, \Delta_0]}$, the computation of (7) can be realized by simply looking up a value (which depends on $\hat{\mu}_0$, $\hat{\mu}_1$, $\hat{k}_0$) in the table of the (cumulative) standard normal distribution [9]. Let us also mention a further possibility of avoiding integration, namely that of basing the termination rule not on the *expected* but on the *most likely* utility $U(\hat{\mu}_1 - \hat{\mu}_0)$.

7 Concluding Remarks

We have proposed an iterative deepening strategy in which the decision whether to terminate or continue search along a path is based on properties of the sequence of node evaluations along that path. The basic idea is to "search deep" in the first place, due to reasons of efficiency, but only as long as the expected utility of the solution that will eventually be found is acceptable. This (heuristic) search principle has been formalized based on statistical methods of change detection. The corresponding search algorithm is parametrized by means of a utility function and a sequence of utility thresholds.

Taking the idea of a *local* termination rule for granted, the criteria (6) and (7) appear natural in the sense that they *fully* exploit the information which has become available

---

[6] This distribution often applies at least approximately.

while following a search path. Observe that admissible (but non-monotone) estimations $f(\eta_k)$ along a path can be aggregated by taking their maximum, which is an obvious way of avoiding a loss of information. As opposed to this, a simple combination (such as, e.g., the average) of all evaluations does hardly make sense in our case since the random variables $F(\eta_k)$ might have different expected values.

The possibility of an efficient parallel implementation served as a main motivation for the development of IDCD and UBID. Indeed, these algorithms are particularly suitable for parallel computation methods since (6) and (7) work with information provided by an individual search path alone and, hence, can lead to a substantial reduction of communication costs. Still, one might think of extending these termination rules by including information from other search paths. This way, it would be possible to base a termination decision on both, the relative change of estimated costs and their absolute values. Particularly, global information can be used for deriving better estimations of the cost of an optimal solution (μ_0 in (5)) and, hence, for improving the test statistic (6). In fact, the methods proposed in this paper can be generalized in further directions as well. Prior information about change points t_0 or deviations Δ, for instance, can be incorporated by means of Bayesian methods or constrained (ML) estimations. Besides, our termination criteria, or variations thereof, might also be useful as stopping rules for selective search in game playing.

References

1. M. Basseville and I.V. Nikiforov. *Detection of Abrupt Changes*. Prentice Hall, 1993.
2. D. Basu. On the elimination of nuisance parameters. *Journal of the American Statistical Association*, 72:355–366, 1977.
3. I. Durdanovic, H. Kleine Büning, and M. Suermann. New aspects and applications in the field of resource-based configuration. Technical Report tr-rsfb-96-023, Department of Computer Science, University of Paderborn, 1996.
4. J. Gaschnig. A problem similarity approach to devising heuristics: First results. In *Proceedings IJCAI-79, 6th International Joint Conference on Artificial Intelligence*, pages 301–307, Tokyo, 1979.
5. O. Hansson and A. Mayer. Heuristic search as evidential reasoning. In *Proceedings 5th Workshop on Uncertainty in AI*, pages 152–161, Windsdor, Ontario, 1989. Morgan Kaufmann.
6. L.R. Harris. The heuristic search under conditions of error. *Artificial Intelligence*, 5(3):217–234, 1974.
7. M. Heinrich. Ressourcenorientiertes Konfigurieren. *Künstliche Intelligenz*, 1/93:11–14, 1993.
8. E.J. Horvitz. Reasoning under varying and uncertain resource constraints. In *Proceeedings AAAI-88*, St. Paul, MN, 1988.
9. E. Hüllermeier. Approximating cost functions in resource-based configuration. Technical Report tr-rsfb-98-060, Department of Mathematics and Computer Science, University of Paderborn, September 1998.
10. E. Hüllermeier and C. Zimmermann. A two-phase search method for solving configuration problems. Technical Report tr-rsfb-98-062, Department of Mathematics and Computer Science, University of Paderborn, October 1998.

11. E. Hüllermeier and C. Zimmermann. A two-phase search method for solving configuration problems. Technical Report tr-rsfb-98-062, Department of Mathematics and Computer Science, University of Paderborn, October 1998.

12. R.E. Korf. Depth-first iterative deepening: An optimal admissible tree search. *Artificial Intelligence*, 27(1):97–109, 1985.

13. G. Lorden. Procedures for reacting to a change in distribution. *Annals of Mathematical Statistics*, 42:1897–1908, 1971.

14. E.S. Page. Continuous inspection schemes. *Biometrica*, 41:100–115, 1954.

15. J. Pearl. *Heuristics: Intelligent Search Strategies for Computer Problem Solving*. Addison-Wesley, 1984.

16. I. Pohl. First results on the effect of error in heuristic search. In B. Meltzer and D. Michie, editors, *Machine Learning 5*, pages 219–236. American Elsevier, New York, 1970.

17. A.E. Prieditis. Machine discovery of effective admissible heuristics. *Machine Learning*, 12:117–141, 1993.

18. A. Wald. *Sequential Analysis*. John Wiley & Sons, New York, 1947.

19. A. Wald and J. Wolfowitz. Optimum character of the sequential probability ratio test. *Annals of Mathematical Statistics*, 19:326–339, 1948.

20. W. Zhang and R.E. Korf. Performance of linear-space search algorithms. *Artificial Intelligence*, 79:241–292, 1995.

Identification of non-additive measures from sample data

Pedro Miranda[1], Michel Grabisch[2] and Pedro Gil[1]

[1] Department of Statistics and Operations Research, University of Oviedo, Spain
[2] Université Paris I- Panthéon- Sorbonne, Paris, France

Abstract. Non-additive measures have become a powerful tool in Decision Making. Therefore, a lot of problems can be solved through the use of Choquet integral with respect to a non-additive measure. Once the decision maker decides to use this criterion in his decision process, next step is to build the non-additive measure up. In this paper we solve the problem of learning the measure from sample data by minimizing the squared error. We study the conditions for the unicity of solution, as well as the set of solutions. A particular family of non-additive measures, the so-called k-additive measures, are specially appealing due to their simplicity and richness. We will use 2-additive measures in a practical case to show that k-additive measures can be considered as a good approximation of general measures.

1 Introduction

Non-additive measures are an important tool in Decision Making. This is due to the fact that they are able to deal with situations of veto and favor, as well as with interaction between criteria (Grabisch (1997a)). Moreover, Choquet Expected Utility Model (Schmeidler (1986)) is able to deal with risk aversion (Chateauneuf (1994)), an important drawback in the Expected Utility Theory (see Allais (1953), Ellsberg (1961)). In recent years, the analysis and use of non-additive measures have been enriched by different equivalent representations of a non-additive measure (Grabisch (1997a)), that are obtained through invertible linear transformations applied on the measure. The most important ones are the Möbius transformation, the Shapley interaction and the Banzhaf interaction. A difficult problem arising in practice has always been the identification of the non-additive measure that, from learning data, best fits these data. If the squared error criterion is considered, then it is known that this problem is equivalent to a quadratic one, but usually solutions found by this procedure appear to be unsatisfactory and rather counter-intuitive.

We believe that performing the optimization through either the Möbius inverse, Shapley interaction or Banzhaf interaction representations should lead to better results. The reason is that these representations are more meaningful for the decision maker than the usual one, i.e. their coefficients are easier to interpret. Consequently, it is easier to extract conclusions about the behaviour of the decision maker.

On the other hand, in order to define a non-additive measure, a set of $2^n - 2$ values must be given. For large values of n, this is a drawback in the approach. k-additive measures appear as a middle term between probability measures and monotone measures in order to find an equilibrium between the richness of monotone measures and the simplicity of probability measures. In this sense, k-additive measures have turned out to be an interesting way to reduce the complexity of monotone measures while keeping good properties from an interpretational point of view. Moreover, as k-additive measures for small values of k (2 or 3) are much simpler than general measures, it turns out that the results in terms of these special measures should be easier to translate in natural language.

The results in the paper are based on those presented in Miranda and Grabisch (1999). We deal with two different problems: First, we study which conditions must be required on the data so that there is a single non-additive measure that best fits them. Taking into account that data are in some practical situations out of our control, we also study the set of solutions of this approximation problem. Secondly, we apply our results to a real problem for the 2-additive case, comparing the results with those obtained for the general case.

The paper is organized as follows: Section 2 is devoted to basic concepts. In Section 3, we recall the results in Miranda and Grabisch (1999). In Section 4, we turn to the problem of unicity of solution and the set of solutions. In Section 5, we look for the optimal 2-additive measure in a real situation and infer properties about the behaviour of the decision maker. Finally, Section 6 is devoted to conclusions and future research.

2   Basic concepts

In order to be self-contained, we present here the basic concepts used throughout the paper. Consider a set of n criteria X, and let us denote $X = \{x_1, ..., x_n\}$.

Definition 1. (Sugeno (1974)) A (discrete) **non-additive measure** (or **fuzzy measure**) on X is a set function $\mu : \mathcal{P}(X) \mapsto [0, 1]$ satisfying

 (i) $\mu(\emptyset) = 0, \quad \mu(X) = 1$ (boundary conditions).
(ii) $A \subset B \Rightarrow \mu(A) \leq \mu(B)$ (monotonicity).

A fuzzy measure is determined by the $2^n - 2$ coefficients on the proper subsets of X.

There are other set functions that can be used to equivalently represent a fuzzy measure. We will need in this paper the so-called Möbius transformation, the Shapley interaction and the Banzhaf interaction.

Definition 2. (Rota (1964)) Let μ be a fuzzy measure on X. The **Möbius transformation** (or **Möbius inverse** or **Möbius transform**) of μ is defined by

$$m(T) := \sum_{K \subset T} (-1)^{|T \setminus K|} \mu(K), \forall T \subset X.$$

In the Theory of Cooperative Games, the Möbius transform is interpreted as the importance of each subset by itself, without considering its parts. In this sense, this transformation is called *dividend* (Owen (1988)).

Definition 3. (Grabisch (1997a)) Let μ be a fuzzy measure on X. The **Shapley interaction index** of μ is defined by

$$I(T) := \sum_{K \subset X \setminus T} \xi_k^{|T|} \sum_{L \subset T} (-1)^{|T|-|L|} \mu(L \cup K)$$

with $\xi_k^p := \frac{(n-k-p)!k!}{(n-p+1)!}$.

The idea of this index is to represent the average contribution of subset A in all coalitions. It generalizes Shapley value (Shapley (1953)).

Definition 4. (Roubens (1996)) Let μ be a fuzzy measure on X. The **Banzhaf interaction index** among the elements of $A \subset X$, is defined by:

$$J(A) = \frac{1}{2^{n-|A|}} \sum_{B \subset X \backslash A} \sum_{C \subset A} (-1)^{|A|-|B|} \mu(B \cup C).$$

This transformation is a generalization of Banzhaf index (Banzhaf (1965)).

Definition 5. (Grabisch (1996)) A fuzzy measure μ is said to be k-**order additive** or k-**additive** if its Möbius transform vanishes for any $A \subset X$ such that $|A| > k$, and there exists at least one subset A of exactly k elements such that $m(A) \neq 0$.

Definition 6. (Choquet (1953)) The **Choquet integral** of a measurable function $f : X \mapsto \mathbb{R}^+$ with respect to a non-additive measure μ is defined by

$$\mathcal{C}_\mu(f) = (C) \int f d\mu := \int_0^\infty \mu(\{x | f(x) > \alpha\}) d\alpha.$$

When X is finite as it is in our case, the expression reduces to:

$$\mathcal{C}_\mu(f) = (C) \int f d\mu := \sum_{i=1}^n (f_{(i)} - f_{(i-1)}) \mu(A_i),$$

where f_i stands for $f(x_i)$, and where parentheses mean a permutation such that $0 = f_{(0)} \leq f_{(1)} \leq \dots \leq f_{(n)}$ and $A_i = \{x_{(i)}, \dots, x_{(n)}\}$.

Choquet integral in terms of Möbius transform is given by:

Proposition 7. *(Walley (1981)) Let μ be a non-additive measure and m its Möbius representation. Then, the discrete Choquet integral with respect to μ of a mapping f is expressed in terms of m by:*

$$\mathcal{C}_\mu(f) = \sum_{A \subset X} m(A) \bigwedge_{x_i \in A} f_i.$$

The following results characterize the classes of Möbius inverses, Shapley and Banzhaf interactions using the conditions of monotonicity:

Theorem 8. *(Chateauneuf and Jaffray (1989)) A set of 2^n coefficients $m(T)$, $T \subset X$ corresponds to the Möbius representation of a non-additive measure if and only if:*

(i) $m(\emptyset) = 0$, $\sum_{A \subset X} m(A) = 1$.
(ii) $\sum_{x_i \in B \subset A} m(B) \geq 0$ *for all $A \subset X$, for all $x_i \in A$.*

Theorem 9. *(Grabisch (1997a)) A set of 2^n coefficients $I(A)$, $A \subset X$, corresponds to the Shapley interaction representation of a fuzzy measure if and only if*

(i) $\sum_{A \subset X} B_{|A|} I(A) = 0$.

(ii) $\sum_{x_i \in X} I(\{x_i\}) = 1$

(iii) $\sum_{A \subset X \backslash x_i} \beta^{|A|}_{|A \cap B|} I(A \cup \{x_i\}) \geq 0, \quad \forall x_i \in X, \forall B \subset X \backslash \{x_i\}$.

Here, the B_k are the Bernoulli numbers, defined recursively by

$$B_k = -\sum_{l=0}^{k-1} \frac{B_l}{k-l+1} \binom{k}{l},$$

starting from $B_0 = 1$, and β^l_k is given by

$$\beta^l_k := \sum_{j=0}^{k} \binom{k}{j} B_{l-j}.$$

Theorem 10. *(Grabisch and Roubens (1996)) A set of 2^n coefficients $J(A)$, $A \subset X$, corresponds to the Banzhaf interaction representation of a non-additive measure if and only if*

(i) $\sum_{A \subset X} (-\frac{1}{2})^{|A|} J(A) = 0$.

(ii) $\sum_{A \subset X} (\frac{1}{2})^{|A|} J(A) = 1$

(iii) $\sum_{A \subset X \backslash x_i} (\frac{1}{2})^{|A|} (-1)^{|A \cap B|} J(A \cup \{x_i\}) \geq 0, \forall x_i \in X, \forall B \subset X \backslash \{x_i\}$.

We conclude this section recalling that the set of all Möbius transformations (resp. Shapley interaction, Banzhaf interaction) coming from non-additive measures is a bounded set.

Theorem 11. *(Miranda and Grabisch (1999)) Let A be a subset of X. Then for every fuzzy measure μ,*

$$m(A) \leq \sum_{j=0}^{l_{|A|}-1} \binom{|A|}{j} (-1)^{|A|-j+1} = \binom{|A|-1}{l_{|A|}-1},$$

where $l_{|A|}$ is given by:

(i) $l_{|A|} = \dfrac{|A|}{2}$ if $|A| \equiv 0 \,(\mathrm{mod} \quad 4)$.

(ii) $l_{|A|} = \dfrac{|A|-1}{2}$ if $|A| \equiv 1 \,(\mathrm{mod} \quad 4)$.

(iii) $l_{|A|} = \dfrac{|A|}{2} + 1$ if $|A| \equiv 2 \,(\mathrm{mod} \quad 4)$.

(iv) $l_{|A|} = \dfrac{|A|+1}{2}$ if $|A| \equiv 3 \,(\mathrm{mod} \quad 4)$.

On the other hand

$$m(A) \geq \sum_{j=0}^{l_{|A|}} \binom{|A|}{j} (-1)^j = (-1) \binom{|A|-1}{l_{|A|}},$$

where $l_{|A|}$ is given by:

(i) $l_{|A|} = \dfrac{|A|}{2} - 1$ *if* $|A| \equiv 0 (\mathrm{mod} \quad 4)$.

(ii) $l_{|A|} = \dfrac{|A|+1}{2}$ *if* $|A| \equiv 1 (\mathrm{mod} \quad 4)$.

(iii) $l_{|A|} = \dfrac{|A|}{2}$ *if* $|A| = 2 (\mathrm{mod} \quad 4)$.

(iv) $l_{|A|} = \dfrac{|A|-1}{2}$ *if* $|A| \equiv 3 (\mathrm{mod} \quad 4)$.

These bounds also hold for Shapley and Banzhaf interactions.

3 Identification based on learning data

Let us start now the study of the non-additive measure that best represents the available data. We assume that we are given a set of sample data. These data come from l individuals, for which we know the scores over each criterium; let us denote the score of individual i over criterium x_j by $f^i(x_j)$, $j = 1, ..., n$, $i = 1, ..., l$. We also know the overall score for individual i, denoted e_i, $i = 1, ..., l$. These scores can be affected by "noise", i.e. these values are not the exact values but approximated ones. The reason is that the decision maker is not able to assign exact values to his preferences.

We assume that both $f^i(x_j)$ and e_i are numerical values. If this is not the case (for example if we have ordinal information), we assume that these data have been converted into numerical data through an ordinal method. More about the learning of fuzzy measures with ordinal information can be found in Miranda et al. (2002) and Miranda (2002) (Chapter 5).

In this section we recall the results obtained in Miranda and Grabisch (1999) for the learning of non-additive measures. It has been proved in Grabisch and Nicolas (1994) that if we consider the quadratic error criterion the problem is equivalent to the following quadratic problem:

$$
\begin{array}{ll}
\text{minimize} & \tfrac{1}{2}\mathbf{u}^T \mathbf{D}_u \mathbf{u} + \boldsymbol{\Gamma}_u^T \mathbf{u} \\
\text{under the constraint} & \mathbf{A}_u \mathbf{u} + \mathbf{b}_u \geq \mathbf{0}
\end{array}
\tag{1}
$$

where $\mathbf{u}$ is the vector containing the values of $\mu(A)$ for all $A \subset X$ and the constraints are the monotonicity conditions.

However, we are interested in obtaining similar expressions for other alternative representations. There are two reasons for this:

- First, the measure obtained is clearer for the decision maker in terms of these representations than in the general form and thus it is easier to interpret the resulting measure or, if several solutions are possible, to choose the most "appropriated". This makes these representations more appealing.
- Secondly, these representations are more suitable for k-additive measures. In the second part of this paper we look for the k-additive measure that best fits our data. For k-additive measures we have $m(A) = 0$, $\forall A \ s.t. \ |A| > k$, and it can be proved that $I(A) = J(A) = 0$, $\forall A \ s.t. \ |A| > k$ (Grabisch (1997a)). Then, the maximal number of non-null variables is $\sum_{i=1}^{k} \binom{n}{i}$. Despite the fact that we can also obtain the fuzzy measure in terms of μ from

$\sum_{i=1}^{k} \binom{n}{i}$ coefficients, the other values do not vanish and tedious computations are needed. For example, for the 2-additive case, we have (Grabisch (1997b))

$$\mu(A) = \sum_{\{x_i, x_j\} \in A} \mu(\{x_i, x_j\}) - (|A| - 2) \sum_{x_i \in A} \mu(\{x_i\}), \forall A \subset X, |A| \geq 2.$$

Consequently, it seems logical to use these alternative representations.

Our first problem was to develop efficient algorithms for passing from a representation to another. This is done through fractal and cardinality transformations.

Definition 12. (Roubens (1996)) The **fractal transformation** is defined recursively by:

$$\mathbf{F}_{(1)} := \begin{bmatrix} f_1 & f_2 \\ f_3 & f_4 \end{bmatrix}, \ f_i \in \mathbb{R}, \ i = 1, 2, 3, 4.$$

$$\mathbf{F}_{(k)} := \begin{bmatrix} f_1 \mathbf{F}_{(k-1)} & f_2 \mathbf{F}_{(k-1)} \\ f_3 \mathbf{F}_{(k-1)} & f_4 \mathbf{F}_{(k-1)} \end{bmatrix}, \ k = 2, \ldots, n.$$

$\mathbf{F}_{(1)}$ is called the **basic fractal matrix**.

Definition 13. (Grabisch et al. (2000)) The **upper-cardinality transformation** is defined recursively from a sequence of $n + 1$ real numbers $c_0, \ldots, c_n$ by:

$$\mathbf{C}_{(1)} := \begin{bmatrix} c_0 & c_1 \\ 0 & c_0 \end{bmatrix}, \ \mathbf{C}_{(1)}^l := \begin{bmatrix} c_{l-1} & c_l \\ 0 & c_{l-1} \end{bmatrix}, \ l = 2, \ldots, n$$

$$\mathbf{C}_{(2)} := \begin{bmatrix} \mathbf{C}_{(1)}^1 & \mathbf{C}_{(1)}^2 \\ 0 & \mathbf{C}_{(1)}^1 \end{bmatrix}, \ \mathbf{C}_{(2)}^l := \begin{bmatrix} \mathbf{C}_{(1)}^l & \mathbf{C}_{1}^{l+1} \\ 0 & \mathbf{C}_{(1)}^l \end{bmatrix}, \ l = 1, \ldots, n - 1$$

$$\mathbf{C}_{(k)} := \begin{bmatrix} \mathbf{C}_{(k-1)}^1 & \mathbf{C}_{(k-1)}^2 \\ 0 & \mathbf{C}_{(k-1)}^1 \end{bmatrix}, \ \mathbf{C}_{(k)}^l := \begin{bmatrix} \mathbf{C}_{(k-1)}^l & \mathbf{C}_{k-1}^{l+1} \\ 0 & \mathbf{C}_{(k-1)}^l \end{bmatrix}, \ \ k = 2, \ldots, n.$$

Definition 14. (Grabisch et al. (2000)) The **lower-cardinality transformation** is defined recursively from a sequence of $n + 1$ real numbers $c_0, \ldots, c_n$ by:

$$\mathbf{C}_{(1)} := \begin{bmatrix} c_0 & 0 \\ c_1 & c_0 \end{bmatrix}, \ \mathbf{C}_{(1)}^l := \begin{bmatrix} c_{l-1} & 0 \\ c_l & c_{l-1} \end{bmatrix}, \ l = 2, \ldots, n$$

$$\mathbf{C}_{(2)} := \begin{bmatrix} \mathbf{C}_{(1)}^1 & 0 \\ \mathbf{C}_{(1)}^2 & \mathbf{C}_{(1)}^1 \end{bmatrix}, \ \mathbf{C}_{(2)}^l := \begin{bmatrix} \mathbf{C}_{(1)}^l & 0 \\ \mathbf{C}_{1}^{l+1} & \mathbf{C}_{(1)}^l \end{bmatrix}, \ l = 1, \ldots, n - 1$$

$$\mathbf{C}_{(k)} := \begin{bmatrix} \mathbf{C}_{(k-1)}^1 & 0 \\ \mathbf{C}_{(k-1)}^2 & \mathbf{C}_{(k-1)}^1 \end{bmatrix}, \ \mathbf{C}_{(k)}^l := \begin{bmatrix} \mathbf{C}_{(k-1)}^l & 0 \\ \mathbf{C}_{k-1}^{l+1} & \mathbf{C}_{(k-1)}^l \end{bmatrix}, \ \ k = 2, \ldots, n.$$

Now, consider the binary order on the subsets of X. This order is obtained as follows: We consider the natural sequence of integers from 0 to $2^n - 1$ and its binary notation $[0]_2, [1]_2, ..., [2^n]_2$ which is $00...00, 00...01, ..., 111...11$. To any number $[i]_2$ in binary notation corresponds one and only one subset $A \subset X$, whence the order. This subset A is defined by those elements $x_j \in X$ such that the j-th coordinate in $[i]_2$ is 1.

Now, let us define $\mathbf{m}$, $\mathbf{I}$, $\mathbf{J}$ the vectors in $\mathbb{R}^{2^n}$ whose coordinates are given by

$$\mathbf{m}[i] = m(A_i), \quad \mathbf{I}[i] = I(A_i), \quad \mathbf{J}[i] = J(A_i),$$

where A_i is the subset in the i-th position if we consider the binary order.

Now, the following can be proved:

Proposition 15. *(Grabisch et al. (2000)) Consider the binary order on X. Then,*

1. *$\mathbf{m} = \mathbf{C}_n \mathbf{u}$, $\mathbf{u} = \mathbf{C}_n^{-1} \mathbf{m}$, where $\mathbf{C}_n$ and $\mathbf{C}_n^{-1}$ are matrices representing lower-cardinality transformations defined by the sequences*

$$c_i = (-1)^i \text{ and } c_i^{-1} = 1, \, i = 1, ..., n.$$

respectively.

2. *$\mathbf{m} = \mathbf{M}_n \mathbf{u}$, $\mathbf{u} = \mathbf{M}_n^{-1} \mathbf{m}$, where $\mathbf{M}_n$ and $\mathbf{M}_n^{-1}$ are matrices representing fractal transformations defined by the matrices*

$$\mathbf{M}_{(1)} := \begin{bmatrix} 1 & 0 \\ -1 & 1 \end{bmatrix}, \, \mathbf{M}_{(1)}^{-1} := \begin{bmatrix} 1 & 0 \\ 1 & 1 \end{bmatrix},$$

respectively.

3. *$\mathbf{I} = \mathbf{C}_n^I \mathbf{m}$, $\mathbf{J} = \mathbf{C}_n^J \mathbf{m}$, where $\mathbf{C}_n^I$ and $\mathbf{C}_n^J$ are matrices representing upper-cardinality transformations defined by the sequences*

$$c_{I,i} = \frac{1}{i+1} \text{ and } c_{J,i} = \frac{1}{2^{i+1}}, \, i = 1, ..., n,$$

respectively. Inverse transformations exist and are also upper-cardinality transformations whose coefficients can be found recursively through the formula

$$c_k^{-1} = -\frac{1}{c_0} \sum_{l=0}^{k-1} \binom{k}{l} c_{k-l} c_l^{-1}, \, k = 1, ..., n.$$

Remark that both cardinality and fractal transformations are $2^n \times 2^n$ matrices and therefore they are not suitable for an efficient implementation. However, in Miranda and Grabisch (1999), we developed an algorithm to obtain the value of the matrix entry given its coordinates. For example, for the upper-cardinality transformation, let us denote by A_i the i-th subset of X in the binary order, and let $\mathbf{y}, \mu$ denote the final and the initial vectors, respectively; $\mathbf{y}$ is initialized $\mathbf{0}$. Then, the algorithm is:

for (i=1 to i=2^n)
    do {

$$y(A_i) = c_0 \cdot \mu(A_i)$$
$$\textbf{for } (j=i+1 \textbf{ to } j=2^n)$$
$$\quad \textbf{do } \{$$
$$\qquad \textbf{if}(A_i \subset A_j)$$
$$\qquad\quad y(A_i) = y(A_i) + c_{|A_i|-|A_j|} \cdot \mu(A_j)$$
$$\qquad \}$$
$$\}$$

The following proposition expresses our problem in terms of Möbius inverse, Shapley interaction and Banzhaf interaction:

Proposition 16. *(Miranda and Grabisch (1999))*

- *The problem of learning a fuzzy measure in terms of its Möbius inverse can be reduced to a quadratic problem with $2^n - 1$ variables (for every Möbius transformation coming from a fuzzy measure we have $m(\emptyset) = 0$) and $n2^{n-1} + 1$ constraints, which can be written:*

$$\begin{aligned} minimize \qquad & \tfrac{1}{2}\mathbf{m}^T\mathbf{D}_m\mathbf{m} + \mathbf{\Gamma}_m^T\mathbf{m} \\ under\ the\ constraints \qquad & \mathbf{f}^T\mathbf{m} = 1 \\ & \mathbf{A}_m\mathbf{m} + \mathbf{b}_m \geq 0 \end{aligned}$$

  where $\mathbf{f}$ is the vector $(1...1)^T$ corresponding to the restriction $\sum_{A \subset X} m(A) = 1$, and $\mathbf{A}_m$ and $\mathbf{b}_m$ correspond to the other constraints of Theorem 8.
- *The problem of learning a fuzzy measure in terms of its Shapley interaction index can be reduced to a quadratic problem with 2^n variables and $n2^{n-1} + 2$ constraints, which can be written:*

$$\begin{aligned} minimize \qquad & \tfrac{1}{2}\mathbf{I}^T\mathbf{D}_I\mathbf{I} + \mathbf{\Gamma}_I^T\mathbf{I} \\ under\ the\ constraints \qquad & \mathbf{f}_1^T\mathbf{I} = 0 \\ & \mathbf{f}_2^T\mathbf{I} = 1 \\ & \mathbf{A}_I\mathbf{I} + \mathbf{b}_I \geq 0 \end{aligned}$$

  where $\mathbf{f}_1$ and $\mathbf{f}_2$ are the vectors corresponding to the two first constraints in Theorem 9, and $\mathbf{A}_I$ and $\mathbf{b}_I$ correspond to the other constraints of Theorem 9.
- *The problem of learning a fuzzy measure in terms of its Banzhaf interaction index can be reduced to a quadratic problem with 2^n variables and $n2^{n-1} + 2$ constraints, which can be written:*

$$\begin{aligned} minimize \qquad & \tfrac{1}{2}\mathbf{J}^T\mathbf{D}_J\mathbf{J} + \mathbf{\Gamma}_J^T\mathbf{J} \\ under\ the\ constraints \qquad & \mathbf{f}_1^T\mathbf{J} = 0 \\ & \mathbf{f}_2^T\mathbf{J} = 1 \\ & \mathbf{A}_J\mathbf{J} + \mathbf{b}_J \geq 0 \end{aligned}$$

  where $\mathbf{f}_1$ and $\mathbf{f}_2$ are the vectors corresponding to two first constraints in Theorem 10, and $\mathbf{A}_J$ and $\mathbf{b}_J$ correspond to the other constraints of Theorem 10.

4   Set of solutions and conditions for unicity

The algorithms explained in Proposition 16 can be used to obtain a solution for the learning problem. However, we have the problem of unicity:

Example 17. Consider $|X| = 3$ and $l = 1$. Moreover, assume that our datum is given by the vector of evaluations $(f^1(x_1), f^1(x_2), f^1(x_3)) = (1, 1, 0)$ and the overall score is $e_1 = 0.5$.

With this datum, $C_\mu(f^1(x_1), f^1(x_2), f^1(x_3)) = \mu(x_1, x_2)$. Then, any measure such that $\mu(x_1, x_2) = 0.5$ is optimal (it fits perfectly our datum!). Therefore, we have an infinite number of solutions. $\Diamond$

It can be argued that there exist multiple solutions because we have not enough data. However, this is not true as next example shows:

Example 18. Consider $|X| = 3$ and $l = 2$. Suppose our data are given by the vectors of evaluations $(f^1(x_1), f^1(x_2), f^1(x_3)) = (1, 0, 0), (f^2(x_1), f^2(x_2), f^2(x_3)) = (0, 1, 1)$ and the overall scores are $e_1 = 1, e_2 = 0$, respectively.

With these data, $C_\mu(f^1(x_1), f^1(x_2), f^1(x_3)) = \mu(x_1)$ and $C_\mu(f^1(x_1), f^1(x_2), f^1(x_3)) = \mu(x_2, x_3)$. Then, the only optimal measure is given by:

$$m(x_1) = 1, \ m(A) = 0 \text{ otherwise.} \ \Diamond$$

Let us then turn to conditions for obtaining a unique solution. We start establishing two counter-intuitive examples to show the difficulties we have to face. First, we may think that if the data are such that all subsets of X are used we will have only one solution. In this case we would only need $\binom{n}{\frac{n}{2}}$ data vectors if n is an even number and $\binom{n}{\frac{n-1}{2}}$ if n is an odd number (Grabisch and Nicolas (1994)). However, this is not true as shown in the following example:

Example 19. Suppose $|X| = 3$ and let us consider a sample of 3 individuals, whose corresponding vectors of evaluations are given by $(1, 0.5, 0), (0.5, 0, 1), (0, 1, 0.5)$. With these data, it is straightforward to see that all subsets are used when computing Choquet integral. Now, if we take as overall scores $e_i = 0.5$, $\forall i$, we will obtain an optimal measure when

$$\mu(A) = 0.5, \forall A \subset X, A \neq X, \emptyset,$$

but another optimal measure is defined by

$$\mu(\{x_1\}) = \mu(\{x_2\} = \mu(\{x_3\}) = 0, \mu(\{x_1, x_2\}) = \mu(\{x_1, x_3\}) = \mu(\{x_2, x_3\}) = 1.$$

Consequently, we have several solutions, even if all subsets have been used. $\Diamond$

On the other hand, we may think that if we have enough data to explore all possible paths from $\emptyset$ to X, i.e. enough data to explore all possible expressions for Choquet integral (at least $n!$ sample data), we will have a unique solution. Remark that this implies the knowledge of a lot of data. However, even in this case, we can obtain several solutions, as it is shown in our next example.

Example 20. Let us take $|X| = 3$ and consider the six data vectors given by

$$(1, 0.5, 0), (1, 0, 0.5), (0.5, 1, 0), (0, 1, 0.5), (0.5, 0, 1), (0, 0.5, 1).$$

It is straightforward to see that all possible orderings are involved. Now, if we take as overall score $e_i = 0.5$, $\forall i$, we will obtain an optimum for

$$\mu(A) = 0.5, \forall A \subset X, A \neq X, \emptyset,$$

but another optimal measure is defined by

$$\mu(\{x_1\} = \mu(\{x_2\} = \mu(\{x_3\} = 0, \mu(\{x_1, x_2\}) = \mu(\{x_1, x_3\}) = \mu(\{x_2, x_3\}) = 1.$$

Thus, we have not a single solution. $\Diamond$

In summary, we conclude that there is no direct relationship between the possible paths and the number of solutions. We have seen that we can get only one solution with two data vectors (Example 18); however, this is a very special case where the values of Choquet integral are 0 or 1 and therefore they lead to a very extreme fuzzy measure.

Then, it seems important to study in depth whether we obtain a unique solution or not from some sample data. We will work with the Möbius inverse of general fuzzy measures. The results obtained also hold for Shapley interaction and Banzhaf interaction and it is straightforward to translate them for the k-additive case.

We start our study with some previous results:

Proposition 21. *Grabisch (1997b) Any convex combination of fuzzy measures is a fuzzy measure, and the four representations are invariant under convex combinations. Formally, let $\mu^1, ..., \mu^p$ be a family of fuzzy measures, with $m^j, I^j, J^j, j = 1, ..., p$ their Möbius, Shapley and Banzhaf representations, and $\alpha_1, ..., \alpha_p \in [0, 1]$ so that $\sum_{i=1}^{p} \alpha_i = 1$. Then, $\mu = \sum_{i=1}^{p} \alpha_i \mu^i$ is a fuzzy measure whose Möbius, Shapley and Banzhaf interactions are given by*

$$m = \sum_{i=1}^{p} \alpha_i m^i, \ I = \sum_{i=1}^{p} \alpha_i I^i, \ J = \sum_{i=1}^{p} \alpha_i J^i.$$

This result is due to the linearity of the representations. As an immediate corollary we have

Corollary 22. *Any convex combination of k-additive measures is another k-additive measure (at most).*

Let us now see the structure of the set of Möbius transforms, Shapley and Banzhaf interactions of fuzzy measures. This will help us later to determine the structure of the set of solutions.

Lemma 23. *Let us denote by $\mathcal{M}$ the set of Möbius transformations (resp. $\mathcal{I}, \mathcal{J}$ the set of Shapley and Banzhaf interactions) of fuzzy measures on X. Then, $\mathcal{M}$ (resp. $\mathcal{I}, \mathcal{J}$) is a convex compact set in the usual topology of $\mathbb{R}^{2^n-1}$ (resp. $\mathbb{R}^{2^n}$)[1]*

Proof: We have to prove that $\mathcal{M}$ (resp. $\mathcal{I}, \mathcal{J}$) is a closed bounded and convex set. For checking that $\mathcal{M}$ (resp. $\mathcal{I}, \mathcal{J}$) is a closed set, it suffices to remark that all monotonicity constraints for these three transformations (Theorems 8, 9 and 10) determine a closed subset of $\mathbb{R}^{2^n-1}$ (resp. $\mathbb{R}^{2^n}$), and that the intersection of closed sets is also a closed set. These sets are bounded by Theorem 11. Finally, they are convex by Proposition 21. ∎

The set of k-additive measures can be proved to be a compact convex set, too.

---

[1] Indeed, the number of independent variables is $2^n - 2$ for these three transformations. However, we have preferred to consider the number of all possible non-null variables.

Suppose that we are given l values of Choquet integral $e_1,...,e_l$ and suppose also that we have the evaluations $f^1(x_1),...,f^1(x_n),...,f^l(x_1),...,f^l(x_n)$. With these values we build the matrix Δ_m, (which is a $l \times (2^n - 1)$ matrix) defined by

$$\Delta_m[i,j] = \min_{i_k \in A_j} \{f^i(x_k)\},$$

where A_j is the subset which is in the j-th position if we consider the binary order in $\mathcal{P}(X)$. Then,

$$\mathcal{C}(f^i(x_1),...,f^i(x_n)) = [\Delta_m]_i \mathbf{m},$$

with $[\Delta_m]_i$ the i-th row of Δ_m.

In our problem we are looking for a Möbius transform m_e such that

$$d(\Delta_m \mathbf{m_e}, \mathbf{e}) = \inf_{m \in \mathcal{M}} \{d(\Delta_m \mathbf{m}, \mathbf{e})\},$$

with $\mathbf{e} = (e_1...e_l)$ the vector of overall scores and d the Euclidean distance (as we are considering the quadratic error criterium). Applying classical results in Topology we have:

Proposition 24. *There is one and only one* $\mathbf{z}$ *in* $\Delta_m \mathcal{M}$ *satisfying*

$$d(\mathbf{z}, \mathbf{e}) = \min_{m \in \mathcal{M}} \{d(\Delta_m \mathbf{m}, \mathbf{e})\}.$$

Proof: Just remark that $\Delta_m \mathcal{M}$ is a closed convex set because so is $\mathcal{M}$. Then, there exists one and only one point $\mathbf{z} \in \Delta_m \mathcal{M}$ satisfying $d(\mathbf{z}, \mathbf{e}) = \min_{m \in \mathcal{M}} \{d(\Delta_m \mathbf{m}, \mathbf{e})\}$ (see e.g. Willard (1970)). ∎

Then, our set of solutions is $\{m \in \mathcal{M} \mid \Delta_m \mathbf{m} = \mathbf{z}\}$. Once $\mathbf{z}$ is given, the problem of unicity reduces to determine whether the system $\Delta_m \mathbf{m} = \mathbf{z}$ under the constraint $m \in \mathcal{M}$ has only one solution.

Lemma 25. *Any convex combination of solutions is another solution. Consequently, the number of solutions is either one or infinity.*

Proof: Let us consider m_1, m_2 the Möbius transforms of two different solutions and take $\lambda \in [0,1]$. By Proposition 21, $\lambda m_1 + (1 - \lambda)m_2 \in \mathcal{M}$. Besides,

$$\Delta_m \mathbf{m_1} = \mathbf{z}, \ \Delta_m \mathbf{m_2} = \mathbf{z} \Rightarrow \Delta_m(\lambda \mathbf{m_1} + (1 - \lambda \mathbf{m_2})) = \lambda \Delta_m \mathbf{m_1} + (1 - \lambda)\Delta_m \mathbf{m_2} = \mathbf{z},$$

whence the result. ∎

Let us start studying the rank of Δ_m, that we will denote by $r(\Delta_m)$. Of course, $r(\Delta_m) \leq 2^n - 1$ (the number of variables). We can easily find examples for which $r(\Delta_m)$ reaches this value:

Example 26. Consider the data given by $f_A(x_i) = 1$ if $x_i \in A$ and $f_A(x_i) = 0$ otherwise, $\forall A \subset X, A \neq \emptyset$. Then we have $l = 2^n - 1$. With these data we will obtain a matrix Δ_m such that $r(\Delta_m) = 2^n - 1$, because considering the data ordered through the binary order, Δ_m is a triangular matrix with $\Delta_m[i,i] = 1$, $\forall i$. $\Diamond$

We are going to study two different cases: First, we study the case when $r(\Delta_m) = 2^n - 1$ and secondly, we deal with the case $r(\Delta_m) < 2^n - 1$.

Proposition 27. *When data information is such that $r(\Delta_m) = 2^n - 1$, there exists one and only one Möbius inverse m_e of a fuzzy measure such that $d(\Delta_m m_e, e) = \min_{m \in \mathcal{M}}\{d(\Delta_m m, e)\}$.*

Proof: Note that at least there exists a solution from Proposition 24. Whenever $r(\Delta_m) = 2^n - 1$ we can transform the system $\Delta_m m = z$ into an equivalent system $\Lambda_m m = z$, where Λ_m is a nonsingular matrix. Thus we have only one solution: $m = \Lambda_m^{-1} z$.   ∎

This result provides us with a sufficient condition for the existence of a unique solution. However, $r(\Delta_m) = 2^n - 1$ implies that we have at least $2^n - 1$ data and this is a very demanding condition on our model, as this number grows exponentially with n. Nevertheless, this is not a necessary condition as we have seen in Example 18, where we showed that a single solution can be found just with two data.

What will happen if $r(\Delta_m) < 2^n - 1$? In this case, the following can be proved:

Theorem 28. *Assume that $r(\Delta_m) < 2^n - 1$, and let m_z be a solution, $\mathbf{v}_1, ..., \mathbf{v}_p$ a basis of $Ker(\Delta_m)$. Then, the set of solutions is given by*

$$\{m_z + \sum_{i=1}^{p} \lambda_i \mathbf{v}_i\},$$

where $\lambda_1, ..., \lambda_p \in \mathbb{R}$ must satisfy the following constraints:

$$\sum_{j=1}^{p} \lambda_j \sum_{B \subset X, B \neq \emptyset} \mathbf{v}_i(B) = 0.$$

$$\sum_{j=1}^{p} \lambda_j \sum_{x_i \in B \subset A} \mathbf{v}_i(B) \geq - \sum_{x_i \in B \subset A} m_z(B), \forall B \subset X, \forall x_i \in B.$$

Proof: Let m_z be a Möbius transform which is a solution of $\Delta_m m = z$. Given another solution $m \in \mathcal{M}$, it is

$$\Delta_m m = z = \Delta_m m_z \Leftrightarrow \Delta_m (m - m_z) = 0 \Leftrightarrow m - m_z \in Ker(\Delta_m).$$

Then, if m is another solution, it can be written as $m = m_z + \mathbf{v}$ with $\mathbf{v} \in Ker(\Delta_m)$.
Now we have to check that m is indeed a Möbius transform of a fuzzy measure.
Let $\{\mathbf{v}_1, ..., \mathbf{v}_p\}$ be a basis of $Ker(\Delta_m)$. Then, if m is another solution, it can be written as $m = m_z + \sum_{i=1}^{p} \lambda_i \mathbf{v}_i$.
The first constraint in Theorem 8 implies:

$$\sum_{B \subset X, B \neq \emptyset} m(B) = 1 \Leftrightarrow \sum_{B \subset X, B \neq \emptyset} m_z(B) + \sum_{i=1}^{p} \lambda_i \sum_{B \subset X, B \neq \emptyset} \mathbf{v}_i(B) = 1 \Leftrightarrow \sum_{i=1}^{p} \lambda_i \sum_{B \subset X, B \neq \emptyset} \mathbf{v}_i(B) = 0.$$

Let us take another constraint, say $\sum_{x_i \in B \subset A} m(B) \geq 0$. This constraint holds if and only if

$$\sum_{i=1}^{p} \lambda_i \sum_{x_i \in B \cap A} \mathbf{v}_i(A) \geq - \sum_{x \in B \cap A} m_z(B),$$

whence we deduce the result.   ∎

Note that we have a single solution if and only if $Ker(\Delta_m) = \emptyset$ (i.e., when $r(\Delta_m) = 2^n - 1$) or $Ker(\Delta_m) \neq \emptyset$ and $\lambda_i = 0$, $\forall i$ is the only solution of the above system ($\lambda_i = 0$, $\forall i$ is always a solution as it leads to m_z). Then, in order to show that we have a unique solution, it suffices to show that the only possible value for λ_i is zero. The first constraint in Theorem 28 implies

$$\sum_{i=1}^{p} \lambda_i \sum_{B \subset X, B \neq \emptyset} \mathbf{v}_i(B) = 0.$$

Let us take another constraint, say $\sum_{x_i \in B \subset A} m(B) \geq 0$. If $\sum_{x_i \in B \subset A} m(B) > 0$, we will always be able to find coefficients λ_i small enough to satisfy this constraint (and hence there is an infinite number of different solutions).

Assume now $\sum_{x_i \in B \subset A} m(B) = 0$. In this case we will need $\sum_{i=1}^{p} \lambda_i \sum_{x_i \in B \subset A} \mathbf{v}_i(A) \geq 0$. Therefore, for determining the unicity, we only need to solve the following problem:

$$\begin{array}{ll} \text{maximize} & \sum_{i=1}^{p} |\lambda_i| \\ \text{under the constraints} & \sum_{i=1}^{p} \lambda_i \sum_{B \subset X, B \neq \emptyset} \mathbf{v}_i(B) = 0 \\ & \sum_{i=1}^{p} \lambda_i \sum_{x_j \in B \subset A} \mathbf{v}_i(B) \geq 0 \end{array} \qquad (2)$$

for any A such that $\sum_{x_j \in B \subset A} m_z(B) = 0$.

We have considered the absolute value because this problem can be transformed in a linear problem.

Note that this process can be also applied to fuzzy measures, Shapley and Banzhaf interaction and k-additive measures because in all these situations the hypotheses (convexity and closed set) hold.

Let us now continue with the study of properties of the set of solutions. Suppose that we have applied the algorithm and that we have found that there are several solutions. Then, we have infinite solutions by Lemma 25. Now, as a consequence of Theorem 28, we have

Corollary 29. *The set of solutions is a polyhedron.*

It is therefore interesting to find the extreme points of this polyhedron. We have seen that the set of solutions is

$$\{m \in \mathcal{M} \mid m = m_z + \mathbf{v}, \ \mathbf{v} \in Ker(\Delta_m), \ \Delta_m m_z = \mathbf{z}\}.$$

If $m \neq m_z$ is another solution, it must satisfy the monotonicity constraints of Möbius inverse. Let us consider $\mathbf{v}_1, ..., \mathbf{v}_p$ a basis of $Ker(\Delta_m)$. Then, $m = m_z + \sum_{i=1}^{p} \lambda_i \mathbf{v}_i$ is a solution if and only if

$$\mathbf{A}\left(m_z + \sum_{i=1}^{p} \lambda_i \mathbf{v}_i\right) \geq \mathbf{b} \Leftrightarrow \mathbf{A}m_z + \mathbf{A}\sum_{i=1}^{p} \lambda_i \mathbf{v}_i \geq \mathbf{b} \Leftrightarrow \sum_{i=1}^{p} \lambda_i \mathbf{A}\mathbf{v}_i \geq \mathbf{b} - \mathbf{A}m_z \Leftrightarrow \sum_{i=1}^{p} \lambda_i \mathbf{w}_i \geq \mathbf{b}',$$

where we have included the first constraint in Theorem 8. This can be written as

$$\mathbf{W}\lambda \geq \mathbf{b}'. \tag{3}$$

Consequently, all solutions are characterized by vector $(\lambda_1, ..., \lambda_p)$. We want to study which are the extreme points of $(\lambda_1, ..., \lambda_p)$ satisfying System (3). Our next result gives necessary and sufficient conditions for $(\lambda_1, ..., \lambda_p)$ to determine an extreme point.

Theorem 30. *Consider* $(\lambda_1, ..., \lambda_p)$ *and let us denote by* $\mathbf{B}$ *the submatrix of* $\mathbf{W}$ *given by those equations that* $(\lambda_1, ..., \lambda_p)$ *satisfies with equality in System 3. Then* $(\lambda_1, ..., \lambda_p)$ *determines an extreme point of the set of solutions if and only if* $(\lambda_1, ..., \lambda_p)$ *satisfies all constraints of System 3 and* $r(\mathbf{B}) = p$.

Proof: Before the proof, note that the first equation of System 3 corresponds to the first constraint of Möbius transform and thus always belongs to $\mathbf{B}$.

Let $\lambda = (\lambda_1, ..., \lambda_p)$ be a solution and let us denote by $\mathbf{B}$ and β the submatrices of $\mathbf{W}$ and $\mathbf{b}$ such that $\mathbf{B}\lambda = \beta$, and by $\mathbf{D}$ and β' the submatrices of $\mathbf{W}$ and $\mathbf{b}$ such that $\mathbf{D}\lambda > \beta'$. Assume now that $r(\mathbf{B}) < p$, i.e. λ does not satisfy with equality at least p linearly independent constraints; we will show that in this case, λ is not an extreme point.

We define

$$\mathcal{S} = \{\lambda' \in \mathbb{R}^p \mid \mathbf{B}\lambda' = \beta\}.$$

Let us prove that $\mathcal{S} \neq \{\lambda\}$: As $r(\mathbf{B}) < p$, there exists $0 \neq \mathbf{x} \in Ker\mathbf{B}$. Now, consider $\lambda + k\mathbf{x}$; then,

$$\mathbf{B}(\lambda + k\mathbf{x}) = \mathbf{B}\lambda + k\mathbf{B}\mathbf{x} = \beta, \ \forall k \in \mathbb{R}.$$

Let us then consider the other monotonicity constraints that are not satisfied with equality by λ, i.e. $\mathbf{D}\lambda > \beta'$. In order for $\mathbf{m}_z + \sum_{i=1}^p (\lambda_i + k x_i)\mathbf{v}_i$ to be a solution, it must be

$$\mathbf{D}(\lambda + k\mathbf{x}) \geq \beta' \Leftrightarrow k\mathbf{D}\mathbf{x} \geq \beta' - \mathbf{D}\lambda = \beta^*.$$

Let us define $\mathbf{D}_1$ as the set of constraints in $\mathbf{D}$ such that $\mathbf{D}_1\mathbf{x} \leq 0$, and $\mathbf{D}_2 = \mathbf{D}\backslash\mathbf{D}_1$.

As $\mathbf{D}\lambda > \beta'$ it is $\beta^* < 0$. Then, for each constraint $\mathbf{c}_i$ in $\mathbf{D}_1$, there exists a positive number $s_i \in \mathbb{R}$ such that the constraint holds for any $k \leq s_i$. Let us take $k_1 = \min_{\mathbf{c}_i \in \mathbf{D}_1}\{s_i\}$. On the other hand, for a positive $k \leq k_1$, constraints in $\mathbf{D}_2$ trivially hold for $\lambda + k\mathbf{x}$. Then, for positive $k \leq k_1$, we have $\lambda + k\mathbf{x}$ is a solution.

The same can be done exchanging the roles of D_1 and D_2. In this case, we obtain a negative value k_2 such that $\lambda + k\mathbf{x}$ is a solution for any negative value of $k \geq k_2$.

As a conclusion, we obtain that $\lambda + k\mathbf{x}$ is a solution for any $k \in [k_2, k_1]$ and $0 \in [k_2, k_1]$. Thus, λ does not define an extreme point as it can be put as a convex combination of two other solutions.

Assume now on the other hand that $r(\mathbf{B}) = p$. Then, we can build a singular matrix $\mathbf{C}$, submatrix of $\mathbf{B}$, such that $\mathbf{C}$ can be inverted. Then,

$$\mathbf{C}\lambda = \mathbf{b}^* \Leftrightarrow \lambda = \mathbf{C}^{-1}\mathbf{b}^*,$$

and thus there is only one solution satisfying with equality these constraints.

Consider λ_1, λ_2 such that $\lambda = \alpha\lambda_1 + (1 - \alpha)\lambda_2$. Then,

$$\mathbf{C}(\alpha\lambda_1 + (1 - \alpha)\lambda_2) = \mathbf{b}^*.$$

On the other hand, as λ_1 and λ_2 are solutions, we have that

$$\mathbf{C}\lambda_1 \geq \mathbf{b}^*, \mathbf{C}\lambda_2 \geq \mathbf{b}^*$$

must be satisfied, whence $\mathbf{C}\lambda_1 = \mathbf{b}^*$, $\mathbf{C}\lambda_2 = \mathbf{b}^*$ and thus $\lambda = \lambda_1 = \lambda_2$ as λ is the only solution of the system. Then, the result holds. ∎

Consequently, applying last theorem, all we have to do for obtaining the extreme points of the set of solutions is to solve all systems having a unique solution. As we have $n2^{n-1}$ constraints, the number of potential systems is $\binom{n2^{n-1}}{p-1}$ (the first constraint is always an equality). This is a very large number, but it serves us as an upper bound. In most practical cases, the number of extreme points does not reach this bound. This is due to the following facts:

- Some systems have several solutions, and thus the solutions obtained cannot be taken as extreme points.
- Some systems have only one solution, but this solution does not satisfy other constraints.
- Some extreme points appear in different systems. This happens when λ satisfies with equality more than p constraints and hence it can be derived from different systems.

Then, the real number of extreme points is usually smaller than $\binom{n2^{n-1}}{p-1}$. Let us remark nevertheless that this bound can be reached:

Example 31. Consider $|X| = 2$ and suppose $l = 1$; moreover, assume that our datum is given by the vector $(1, 1)$ and the overall score is $e_1 = 1$. Then, every non-additive measure is a solution as this datum leads to the constraint $\mu(X) = 1$.

Consider the binary order $x_1 \prec x_2 \prec \{x_1, x_2\}$. If we consider the set of all non-additive measures, it is straightforward to see that the extreme points in terms of the Möbius transform are

$$(0, 0, 1), (1, 0, 0), (0, 1, 0), (1, 1, -1).$$

On the other hand, the matrix Δ_m is given by the vector $(1, 1, 1)$ and therefore we have $p = r(Ker(\Delta_m)) = 2$.

Consequently, $\binom{n2^{n-1}}{p-1} = \binom{4}{1} = 4$ and the bound is reached. ◊

5 Experimental results

We are going to study in this section the differences in the results of learning fuzzy measures when we consider general fuzzy measures and the 2-additive case. We study the notion of richness of a cream in the field of cosmetics.

10 products have been measured by 10 judges. We have 8 factors that can have influence in the richness of the cream. These factors are:

factor 1	prehension with finger (Fi)
factor 2	onctuousness (On)
factor 3	spreading (Sp)
factor 4	penetration speed (Pe)
factor 5	stickiness (St)
factor 6	brightness (Br)
factor 7	remanence (Re)
factor 8	greasiness (Gr)

This study has been done in (Grabisch et al. (1997)). In it, they study the problem of identifying the fuzzy measure with the subroutines HLMS (Grabisch (1995)) and QUAD (standard methods of quadratic programming). It is also considered the arithmetic mean of judges opinions for the final evaluation over each criterium. They found an error of 0.122 for HMLS and 0.083 for QUAD. After removing two criteria, namely Br and Fi that are negligible, they obtain the following results for Shapley values (after normalizing × 6, so that a value greater than 1 means a criterion more important than the average), and interaction values for pairs:

```
Shapley values
---------------------------------
         On    : 1.450409
         Sp    : 0.874723
         Pe    : 0.715366
         St    : 0.828624
         Re    : 1.339188
         Gr    : 0.791691


Interaction matrix
---------------------------------------------
(1,2)=-0.101 (1,3)=-0.067 (1,4)=-0.068 (1,5)=-0.021 (1,6)= 0.056
             (2,3)= 0.033 (2,4)=-0.038 (2,5)=-0.108 (2,6)= 0.046
                          (3,4)=-0.083 (3,5)=-0.049 (3,6)= 0.001
                                       (4,5)= 0.075 (4,6)=-0.038
                                                    (5,6)= 0.100
```

The conclusions of the results are:

– The most complement pairs of criteria are (Re, Gr), (St, Re) and (On, Gr).
– The most redundant pairs are (On, Sp), (Sp, Re), (On, Pe) and (On, St).
– Re and On are the most important criteria.

We are going to use the Powel-Schittkowski method of identification, that allows us to consider the quadratic error model with monotonicity constraints. We study the 2-additive case to show that the results are similar to those for the general case, and thus we conclude that k-additive measures can be used for applications instead of general fuzzy measures without incurring in a big error.

We start with all the factors. If we apply the algorithm in this case we obtain the following results:

```
quadratic error : 0.082927

Shapley values
-----------------------------------
  Fi : 0.113010
  On : 1.897706
  Sp : 0.697659
  Pe : 0.000000
  St : 0.000000
  Br : 0.000000
  Re : 5.178023
  Gr : 0.113302
```

```
Interaction matrix
-----------------------------------
(1,2)=0.00  (1,3)= 0.03  (1,4)=0.00  (1,5)=0.00  (1,6)=0.00  (1,7)= 0.00  (1,8)=0.00
            (2,3)= 0.03  (2,4)=0.00  (2,5)=0.00  (2,6)=0.00  (2,7)=-0.45  (2,8)=0.00
                         (3,4)=0.00  (3,5)=0.00  (3,6)=0.00  (3,7)=-0.09  (3,8)=0.03
                                     (4,5)=0.00  (4,6)=0.00  (4,7)= 0.00  (4,8)=0.00
                                                 (5,6)=0.00  (5,7)= 0.00  (5,8)=0.00
                                                             (6,7)= 0.00  (6,8)=0.00
                                                                          (7,8)=0.00
```

It can be observed that there are three factors that have no importance, namely the factors Pe, St and Br. On the other hand, the quadratic error is rather small and, therefore, we conclude that we have obtained a good model.

Then, we remove the negligible criteria and repeat the experiment. In this case, we obtain:

```
quadratic error : 0.082927

Shapley values
-----------------------------------
  Fi : 0.070818
  On : 1.186067
  Sp : 0.436037
  Re : 3.236264
  Gr : 0.070813
```

```
Interaction matrix
-----------------------------------
I( 1, 2)=-0.000   I( 1, 3)= 0.028   I( 1, 7)= 0.000   I( 1, 8)=-0.000
                  I( 2, 3)= 0.028   I( 2, 7)=-0.446   I( 2, 8)=-0.000
                                    I( 3, 7)=-0.089   I( 3, 8)= 0.028
                                                      I( 7, 8)=-0.000
```

Finally, as an example of what happens if we delete a non-negligible factor, we remove Gr, that seems to be the less important criterion. In this case, we obtain:

```
quadratic error : 0.082927
```

```
Shapley values
-------------------------------
 Fi : 0.084980
 On : 0.977179
 Sp : 0.348829
 Re : 2.589011

Interaction matrix
---------------------------------------
I( 1, 2)=-0.000   I( 1, 3)= 0.042   I( 1, 7)= 0.000
                  I( 2, 3)= 0.042   I( 2, 7)=-0.446
                                    I( 3, 7)=-0.089
```

We can observe that the results are slightly different. The reason is that factor Gr has a positive Shapley value (in the other cases the deleted factor had Shapley value zero). Thus, it can be obtained another solution as good as the first one removing criterium Gr, but in this case its influence must be shared by other criteria, i.e. influence of criterium Gr can be explained from other criteria.

Note that in all the cases we obtain that the factors On and Re are very important and that they have a negative interaction. Thus, we can conclude that the model is

$$e \approx a_2 \vee a_7,$$

and the most important criterion is a_7 (Re). It is interesting to note that for the 2-additive case, Re is much more important than in the general case. Remark also that these conclusions could be obtained from the results with five criteria.

As a conclusion, we see that we obtain almost the same conclusions as we obtained in the general case. However, we have simplified the number of coefficients used in the computations. This translates in a faster answer. On the other hand, we have reduced the model to a very simple formula, easy to understand by the decision maker, and to extract conclusions and properties.

6 Conclusions

The practical learning of the non-additive measure modeling a decision making problem is a difficult problem. In a previous paper, using the related work of Grabisch and Nicolas, we dealt with the problem of identifying a non-additive measure in terms of other alternative representations, namely the Möbius transformation, the Shapley interaction index and the Banzhaf interaction. This problem was specially interesting when dealing with k-additive measures as we obtained a reduction in the number of coefficients.

However, there exists a problem when several measures fit equally good the sample data. In this paper we continue the work started in (Miranda and Grabisch (1999)). Given a set of sample data, we give conditions for obtaining a single solution. We also propose an algorithm for computing whether we have only one solution. If we have several solutions, the set of all possible solutions is a convex polyhedron; in this work we characterize the extreme points of this set and we give an upper bound of the number of vertices.

Finally, we apply our algorithm in a real situation. This allows us to conclude that k-additive measures are a good approximation of general non-additive measures and thus, due to their reduced complexity, a very appealing tool in Decision Making.

Sometimes, we can use more data than needed in order to ensure unicity. Moreover, it is possible to obtain several solutions due to the "noise":

Example 32. Consider the data from Example 18. In such example, we had a unique solution. Assume that we have a third datum, given by the vector $(f^3(x_1), f^3(x_2), f^3(x_3)) = (1, 0, 0)$ again, but now, due to noise, the overall score is $e_3 = 0.9$. In this case, the optimal values are $\mu(x_2, x_3) = 0$ and $\mu(x_1) = 0.95$. But then, $\mu(1, 2)$ can attain any value in $[0.95, 1]$ and thus unicity is lost. $\Diamond$

A possible solution can be a step-by-step algorithm. Consider a set of data leading to different solutions and add another datum. If the new set of data leads to a single solution, solve the corresponding problem. Otherwise, add another datum and repeat the process. However, due again to "noise" in the data, maybe this solution is not the real measure representing the behaviour of the decision maker. In this sense, a deeper research is needed.

References

Allais, M. (1953). Le comportement de l'homme rationnel devant le risque: critique des postulats de l'école américaine. *Econometrica* (21):503–546. In French.

Banzhaf, J. F. (1965). Weighted voting doesn't work: A mathematical analysis. *Rutgers Law Review* (19):317–343.

Chateauneuf, A., and Jaffray, J.-Y. (1989). Some characterizations of lower probabilities and other monotone capacities through the use of Möbius inversion. *Mathematical Social Sciences* (17):263–283.

Chateauneuf, A. (1994). Modelling attitudes towards uncertainty and risk through the use of Choquet integral. *Annals of Operations Research* (52):3–20.

Choquet, G. (1953). Theory of capacities. *Annales de l'Institut Fourier* (5):131–295.

Ellsberg, D. (1961). Risk, ambiguity, and the Savage axioms. *Quart. J. Econom.* (75):643–669.

Grabisch, M., and Nicolas, J.-M. (1994). Classification by fuzzy integral-performance and tests. *Fuzzy Sets and Systems, Special Issue on Pattern Recognition* (65):255–271.

Grabisch, M., and Roubens, M. (1996). Equivalent representations of a set function with application to decision making. In *6th IEEE Int. Conf. on Fuzzy Systems*.

Grabisch, M., Baret, J.-M., and Larnicol, M. (1997). Analysis of interaction between criteria by fuzzy measures and its application to cosmetics. In *Proceedings of Int. Conf. on Methods and Applications of Multicriteria Decision Making*.

Grabisch, M., Marichal, J.-L., and Roubens, M. (2000). Equivalent representations of a set function with applications to game theory and multicriteria decision making. *Mathematics of Operations Research* 25(2):157–178.

Grabisch, M. (1995). A new algorithm for identifying fuzzy measures and its application to pattern recognition. In *Proceedings of Int. Joint Conf. of the 4th IEEE Int. Conf. on Fuzzy Systems and the 2nd Int. Fuzzy Engineering Sysmposium*, 145–150.

Grabisch, M. (1996). *k*-order additive discrete fuzzy measures. In *Proceedings of 6th Int. Conf. on Information Processing and Management of Uncertainty in Knowledge-Based Systems (IPMU)*, 1345–1350.

Grabisch, M. (1997a). Alternative representations of discrete fuzzy measures for decision making. *Int. J. of Uncertainty, Fuzziness and Knowledge-Based Systems* 5:587–607.

Grabisch, M. (1997b). *k*-order additive discrete fuzzy measures and their representation. *Fuzzy Sets and Systems* (92):167–189.

Miranda, P., and Grabisch, M. (1999). Optimization issues for fuzzy measures. *International Journal of Uncertainty, Fuzziness and Knowledge-Based Systems* 7(6):545–560. Selected papers from IPMU'98.

Miranda, P., Grabisch, M., and Gil, P. (2002). An algorithm for identifying fuzzy measures with ordinal information. In Grzegorzewski, P., Hryniewicz, O., and Gil, M. A., eds., *Soft methods in Probability, Statistics and Data Analysis*, Advances in Soft Computing. Heidelberg (Germany): Physica-Verlag. 321–328.

Miranda, P. (2002). *Applications of k-additive measures to Decision Theory*. Ph.D. Dissertation, University of Oviedo.

Owen, G. (1988). Multilinear extensions of a game. In Roth, A. E., ed., *The Shapley value. Essays in Honor of Lloyd S. Shapley*. Cambridge University Press. 139–151.

Rota, G. C. (1964). On the foundations of combinatorial theory I. Theory of Möbius functions. *Zeitschrift für Wahrscheinlichkeitstheorie und Verwandte Gebiete* (2):340–368.

Roubens, M. (1996). Interaction between criteria and definition of weights in MCDA problems. In *44th Meeting of the European Working Group Multicriteria Aid for Decisions*.

Schmeidler, D. (1986). Integral representation without additivity. *Proc. of the Amer. Math. Soc.* (97(2)):255–261.

Shapley, L. S. (1953). A value for n-person games. In Kuhn, H. W., and Tucker, A. W., eds., *Contributions to the theory of Games*, volume II of *Annals of Mathematics Studies*. Princeton University Press. 307–317.

Sugeno, M. (1974). *Theory of fuzzy integrals and its applications*. Ph.D. Dissertation, Tokyo Institute of Technology.

Walley, P. (1981). Coherent lower (and upper) probabilities. Technical Report 22, University of Warwick, Coventry, (UK).

Willard, S. (1970). *General Topology*. Sidney (Australia): Addison-Wesley Publishing Company.

Notification Planning
with Developing Information States

Markus Schaal[*][†] and Hans-J. Lenz [†‡]

[†] Inst. of Production, Informations Systems, Operations Research, Free University, Berlin
[‡] Inst.of Statistics and Econometrics, Free University, Berlin

Abstract Besides triggered notification, intelligent notification keeps track of human users plans and events to be expected in the future. As states of the notification system vary in time, best choices for notification do as well. In order to employ the knowledge about future information states, notification planning is modelled by influence diagrams with developing information states explicitly given per notification time point.

So far, the approach is restricted to the application domain of route guidance, where human users plans are well structured and information about expected events is available.

1 Introduction

We consider decisions about traveller notification. The general idea is as follows: Let L be the set of all locations. The traveller needs to get from start location $l_{start} \in L$ to destination $l_{dest} \in L$ (another location). He can choose among a set of routes R. Any route $r \in R$ starts from l_{start} and leads to l_{dest}, i.e. $r = \{l_1, \ldots, l_n\}$ is a sequence of locations ($l_i \in L$ for $1 \leq l \leq n$) with $l_1 = l_{start}$ and $l_n = l_{dest}$. Route concatenation is denoted by a dot-notation ($\cdot$), e.g. $(a, b, c, e) = (a, b) \cdot (c, e) = a \cdot b \cdot c \cdot e$.

We employ a running example which is shown in Fig. 1 with routes A, B and C each starting from location a and leading to destination g.

The traveller is to be supported by notification about the optimal route. The optimal route may change in the course of time due to event occurrence (e.g. congestion announcement, train delay, etc.).

The initial route set R is pre-computed by an algorithm for shortest paths, cf. e.g. Lawler (1976).

2 Concepts

The so-called *information state* is used for the prediction of travel time. The information state changes (develops) in the course of time. Empirical analysis of the information state

[*]This research was supported by the German Research Society, Berlin-Brandenburg Graduate School in Distributed Information Systems (DFG grant no. GRK 316).

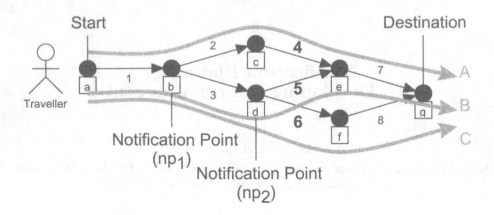

Figure 1. Route Graph for a road network

history (given as state change events) may provide the background for estimation of information state transition probabilities. Let S be the finite set of possible information states. For a given information state $s \in S$ at time t we assume knowledge about the conditional probability of state $s' \in S$ at time $t' \in T$ with $t' > t$. This so-called *transition probability* is denoted by $p_{ss'}(t, t')$.

The information state consists of a sub-state per route segment, i.e. $s = (s_1, \ldots, s_p)$ with s_i for $1 \leq i \leq p$ denoting the sub-state for route segment i. The state space $S = S_{edge}^p = S_{edge} \times \cdots \times S_{edge}$ is the cross product of the local state space per edge with $S_{edge} = \{c, f\}$ ($c \equiv$ "edge is congested", $f \equiv$ "free flow on edge") .

A notification sent to the user is characterized by *time, location, content* and *presentation*, i.e. its information logistics dimensions except the *mode of transmission* which is implicitly given as *push*, see Deiters and Lienemann (2001) as an introduction to information logistics. Finding the best time, location, content and presentation simultaneously is a nearly impossible task.

We apply the *Invariant Embedding Principle*. Therefore we can focus on so-called *notification (decision) points*, i.e. locations where alternatives for route selection exist. Note, that the same location may be reached via different routes, e.g. location e can either be reached via route (a, b, c, e) or by route (a, b, d, e). Therefore, we consider so-called pre-routes instead of locations. $R_{pre} = \{r \mid \exists_{r'} \; r \cdot r' \in R\}$ is the *pre-route set of* R. In other words, any sequence of locations which starts a route in R is a pre-route. For any pre-route $r \in R_{pre}$, $Succ(r)$ denotes the set of successors, i.e. $Succ(r) = \{r' \in R_{pre} \mid \exists_{l \in L} \; r' = r \cdot l\}$. Now, a *notification (decision) point* is a pre-route with more than one successor, i.e. pre-route $r \in R_{pre}$ is a notification point if $\mid Succ(r) \mid > 1$.

Our running example has two notification (decision) points, namely $np_1 = (a, b)$ and $np_2 = (a, b, d)$. Route decisions are taken and notifications are sent at notification points, i.e. a notification point is also a point for replanning the best route and replacement of the route to follow. Therefore, the decision for a route is actually only a decision about a route segment, i.e. the partial route from the current location to the next decision point. In our example, we have alternative partial routes (b, c, e, g) and (b, d) at notification point np_1 and alternative partial routes (d, e, g) and (d, f, g) at notification point np_2. Even though, the traveller is notified about complete routes. Partial routes are completed such that the probability of an update is minimal. A *default route* of the traveller is the route which he has been lastly notified.

Future information states are needed for future decisions. Since we consider the effect of future decisions on the analysis of current decisions, a look-ahead on future information states is needed. We employ a discrete dynamic model due to our *Invariant Embedding Principle*. We consider a finite number of isolated notifications points. Estimated time points and random information states are linked to notification points.

Our utility model is based on the travellers's arrival time only. We distinguish the utility of a concrete arrival time and the utility of a random arrival time. An utility function $u : T \rightarrow \mathbb{R}_0^+$ is an order preserving function of the preferences of the user with respect to a concrete arrival time. Let t_1 and t_2 be arrival time options. Then $u(t_1) < u(t_2)$ iff the user prefers t_2 over t_1. The utility u_t of the random arrival time $\mathbf{t}$ is the expectation of the utility for specific arrival times, i.e. $u_{\mathbf{t}} = E(u(\mathbf{t})) = \sum_{t \in T} u(t) \cdot p(\mathbf{t} = t)$. In fact, this is the only utility function (up to linear transformations) for the random quantity $\mathbf{t}$ which fulfills the four axioms of preference as stated by von Neumann and Morgenstern (1944) ($\mathbf{t}$ is a lottery for t).

For a current notification point r_{curr}, a current time t_{curr} and a current information state s_{curr} a route $r \in R$ is searched which should be followed until the next notification point is reached. The solution is found as the result of an optimization with respect to expected utility. However, route r does not necessarily provide the highest expected utility for the current information state, since future notification options are considered. The latter are based on future information states.

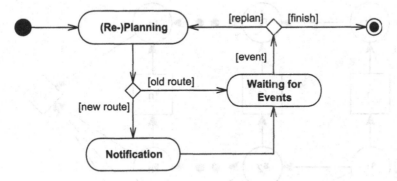

Figure 2. Activity Diagram for Intelligent Notification

Generally, we will track the traveller during his journey and notification will be re-planned while actually approaching a notification point or in the case of new incoming events concerning the information state. This is necessary even though future states have been considered at previous decisions. The reason is, that time points of future events such as reaching a location have only been estimated and updates from actual observation need to be considered. An activity diagram shows the different actions to be

performed for intelligent notification and the events upon which action has to be taken
(cf. Fig. 2).

After initial *planning* (which results in a *notification* about the new default route for
the traveller), the information system is *waiting for events*. Possible events are state
changes in the information state or the location event telling about the traveller's approach to a notification point. On event occurrence, the whole task of intelligent notification is either finished or *replanning* is started.

3 Influence Diagram Representation

Influence diagrams are directed graphs with three types of nodes (cf. Howard and Matheson (1981), Pearl (1991) and Shachter (1987)). Chance nodes (shown as ovals) represent
random variables, decision nodes (shown as rectangles) represent possible decisions and
value nodes (shown as diamonds) represent rewards and costs for decisions and outcomes
of random variables. Nodes are connected by directed arcs. Arcs leading to chance nodes
denote conditional dependency, arcs leading to value nodes denote functional dependency
and arcs leading to decision nodes are informational, i.e. the respective value is known
before a decision has to be made.

Let $R_{np} = \{r \in R_{pre} \mid (r_{curr} \subseteq r) \wedge (r \text{ is a notification point})\}$ be the set of all
notification points to be reached from current pre-route r_{curr}. We consider the case
of $R_{np} = r_1^{np}, \ldots, r_n^{np}$ with $\forall_{1 \leq i < n} r_i^{np} \subset r_{i+1}^{np}$, i.e. all notification points are linearly
ordered.

Furthermore, we assume a fixed sequence of time points $\hat{t}_1, \ldots, \hat{t}_n$, where $\hat{t}_i$ is an
estimate of the arrival at notification point r_i^{np}. The following state and decision nodes
are used:

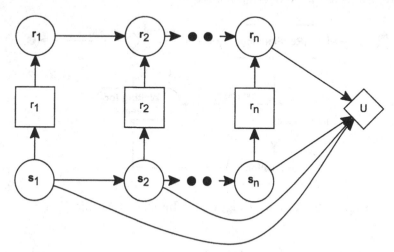

Figure 3. Influence Diagram for Notification Point Sequence $R^{np} = r_1^{np}, \ldots, r_n^{np}$

- $r_i \in R$ is the route decision at notification point r_i^{np}, i.e. $(r_i^{np}, \hat{t}_i, r_i)$ is the selection notification option
- $\mathbf{r}_i \in R$ is the uncertain route selection of the traveller using decision r_i. The route selection at time $\hat{t}_i$ depends upon the the previous route selection and also on the acceptance of route decision r_i by the traveller.
- $\mathbf{s}_i$ is the uncertain information state at the notification point r_i^{np}.

Fig. 3 shows the resulting influence diagram. The utility of a specific set of decisions is given by the expected utility of the arrival time at destination l_{dest}. To compute the random arrival time we need the route (represented by traveller's final route selection $\mathbf{r}_n$) and the sequence of information states $\mathbf{s}_1, \ldots, \mathbf{s}_n$.

Note, that it is not sufficient to consider the final information state $\mathbf{s}_n$ for utility assessment. This is due to the fact that the information state is an abstraction of the complete state necessary for utility assessment.

4 Example

The influence diagram for our running example is shown in figure 4. We give a description of the nodes below.

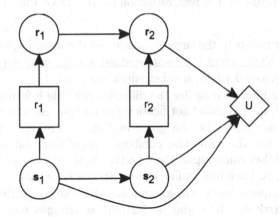

Figure 4. Influence Diagram for the Road Scenario

Chance Nodes:

$\mathbf{r}_1$ $\{A, B, C\}$, the actual route decision after the first notification point np_1;

$\mathbf{r}_2$ $\{A, B, C\}$, the actual route decision after the second notification point np_2;

$\mathbf{s}_1$ $\{(s_1, \ldots, s_8) \mid s_i \in \{c, f\}\}$, the information state at (before) the first notification option;

$\mathbf{s}_2$ $\{(s_1, \ldots, s_8) \mid s_i \in \{c, f\}\}$, the information state at (before) the second notification option;

Decision Nodes:

r_1 $\{A, B, C\}$, the route decision at the first notification option.

r_2 $\{A, B, C\}$, the route decision at the second notification option.

Value Nodes:

U, the utility node.

For the example, a congestion on edges (c, e), (d, e) and (d, f) is depicted in Fig. 5.

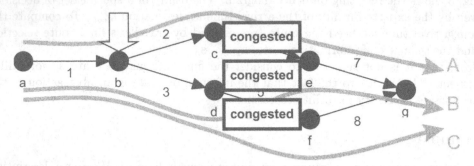

Figure 5. Current Situation on Route Graph

The white big arrow points to the current location of the traveller, i.e. the current pre-route is $r_{curr} = (a, b)$. The current information state is $s_{curr} = \mathbf{s}_1 = (f, f, f, c, c, c, f, f)$, i.e. edges $4, 5, 6$ are congested while all other edges are free.

In a second step, we have to consider the different possible information states at the estimated future time $\hat{t}_2$ of the second notification option np_2. At current time $t_{curr} = \hat{t}_1$, the information state is simply the one given in Fig. 5. For $\hat{t}_2$, only four situations are considered. These are the respective combinations of free and/or congested route segments 5 and 6. All other route segments are either inaccessible (route segments 2 and 4) or they are assumed to have free traffic anyway (route segments 7 and 8).

Probabilities are computed under the assumption that the congestion will vanish with probability 0.8 when arriving there and no further congestions will occur. Thus, the probability of the different information states is given by probabilities $p_1 = 0.8^2 = 0.64$ (both congestions vanished), $p_2 = p_3 = 0.8 \cdot (1 - 0.8) = 0.16$ (one congestion vanished, two cases) and $p_4 = (1 - 0.8)^2 = 0.04$ (no congestion vanished).

With implicit deadline $t_{deadline}$ and random arrival time $\mathbf{t}_g$ at location g, the traveller's utility for arrival time is given by:

$$U(\mathbf{t}_g = t) = \begin{cases} 10 & t \le t_{deadline} \\ 0 & t > t_{deadline} \end{cases} \tag{4.1}$$

However, we omit the stochastic plan evaluation here. Instead, we assume to be in time at location g if the traveller has no congestion on his route, i.e. the utility is zero if the respective sub-state is "congested" and equal to 10, if the respective sub-state is "free" at the time $\hat{t}_2$. For instance, consider route A and road segment 4. Then

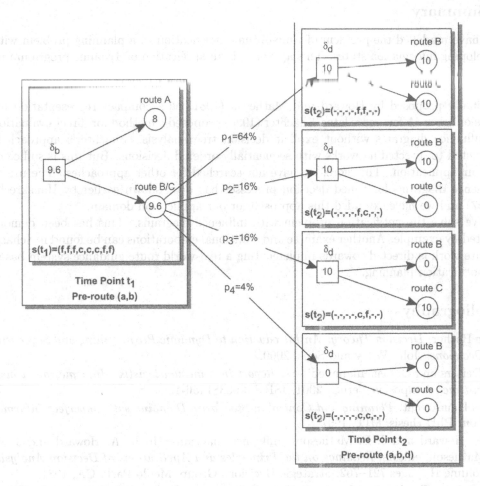

Figure 6. Decision Tree

$$U(\mathbf{r}_d = "A", \mathbf{s}_2 = (*, *, *, x, *, *, *, *)) = \begin{cases} 10 & x = "f" \\ 0 & x = "c" \end{cases} \quad (4.2)$$

These utilities are written in the rightmost circles and propagated to the left by weighting with the respective probabilities (in the case of lotteries) or by selecting the optimal value (in the case of decisions).

Thus, the utility of "waiting" at current time is $U = 9.6$) and is higher than "notifying" at current time, with utility $U = 0.8$. This result is intuitively clear, since waiting provides us with better information for choosing between route B and route C, while taking route A gives exactly the reward corresponding to the probability of one congestion to disappear. Other situations for much more complex route sets R may be solved by this approach as well.

5 Summary

We have modelled the problem of time-optimal notification as a planning problem with developing information states. This approach is an application of dynamic programming and sequential decisions, cf. e.g. Bather (2000).

We represent our planning problem as an influence diagram. Influence diagrams where originally proposed by Howard and Matheson (1981) as a compact representation for decision trees. A few years later, Shachter (1986) proposed a method for direct evaluation of influence diagrams without explicit decision tree analysis. Shachter's approach is restricted to directed networks with sequentially ordered decisions. But this is sufficient for our application. Therefore, we have not searched for other approaches. The use of influence diagrams for timed decision processes has been shown earlier by Hauskrecht (1997) and we have extended this approach for our application domain.

We solve the notification problem with influence diagrams. This has been demonstrated by example. Another example and additional elaborations can be found in Schaal. Future work is directed towards implementing a real-world route guidance system based on notification planning.

Bibliography

John Bather. *Decision Theory: An Introduction to Dynamic Programming and Sequential Decisions*. John Wiley and Sons, 2000.

W. Deiters and C. Lienemann, editors. *Report Informationslogistik - Informationen just-in-time*. Symposion Verlag, 2001. ISBN 3-93381456-1.

Milos Hauskrecht. *Planning and Control in Stochastic Domains with Imperfect Information*. PhD thesis, MIT, 1997.

R. A. Howard and J. E. Matheson. Influence diagrams. In R. A. Howard and J. E. Matheson, editors, *Readings on the Principles and Applications of Decision Analysis*, volume II, pages 721–762. Strategic Decisions Group, Menlo Park, CA, 1981.

E. Lawler. *Combinatorial optimization - networks and matroids*. Holt, Rinehart and Winston, New York, 1976.

Judea Pearl. *Probabilistic Reasoning in Intelligent Systems: Networks of Plausible Inference*. Morgan Kaufmann, 1991. (Revised 2nd Edition).

Markus Schaal. *Time-Optimal Notification under Uncertainty*. PhD thesis, Technische Universtät, Berlin. forthcoming.

Ross D. Shachter. Evaluating Influence Diagrams. *Operations Research*, 33(6), 1986.

Ross D. Shachter. Probabilistic Inference and Influence Diagrams. *Operations Research*, 36(4):589–604, 1987.

J. von Neumann and O. Morgenstern. *Theory of Games and Economic Behaviour*. Princeton University Press, 1944.

SESSION II

PLANNING, CONTOL and LEARNING

The Problem of Planning with Three Sources of Uncertainty

Paolo Traverso

ITC/IRST, Institute for Scientific and Technological Research 38050 - Povo - Trento, Italy

Abstract Planning under uncertainty is one of the most significant and challenging problems in Artificial Intelligence and Computer Science. In this paper, we focus on the following sources of uncertainty: *nondeterminism*, *partial observability*, and *extended goals*. We discuss how the "Planning as Model Checking" approach can deal with these three forms of uncertainty.

1 The Problem: Three Sources of Uncertainty

We focus here on the following sources of uncertainty: nondeterministic models, partial observability, and extended goals. *Nondeterministic models* allow for modeling uncertainty in the outcome of actions. A nondeterministic model can take into account the fact, e.g., that beyond the nominal behaviour, a component may fail, and planners with nondeterministic models should plan for "exception-handling" or recovering mechanisms. Neglecting this source of uncertainty means to assume that the nominal case accounted for by a deterministic model is highly frequent, cases non taken into account being marginal, and/or they can be easily dealt with at the controller level. However, in several applications, non-nominal outcomes of actions are important, e.g., highly critical. They should be modeled at planning time as well as nominal outcomes. For instance, it is very important to model the possible failures of an action that moves a railway switch or that sets the signals of a rail-road crossing. Moreover, in some cases, nominal outcomes do not exist, like when we throw dice or toss a coin. This is the case of actions that ask for information to the user, that query a data base, or that model web services. *Planning with nondeterministic models* leads to some main difficulties. A plan may result in many different execution paths. Planning algorithms need efficient ways to analyze all possible action outcomes and generate plans that have conditional and iterative behaviours.

A second source of uncertainty is *partial observability*. In several realistic problems, the state of the system is only partially visible at run-time and, as a consequence, different states of the system are indistinguishable for the controller. This is the case of the "home sequencing" problem, i.e., the problem of re-initializing a microprocessor by executing some commands without having access to its internal registers. Moreover, in some applications, some variables can be observable just in some states, or after some "sensing actions" have been executed. For instance, a mobile robot in a room may not know whether the door in another room is opened until it moves to the room. A planner that composes web-services for a travel agency cannot know whether there will be seats available until it queries the web service. *Planning under partial observability* has been shown to be a hard problem, both theoretically and experimentally. The main technical consequence of partial observability is that observations return sets of states rather than single states. This makes the search space no longer the set of states of the domain, but its power-set.

A third dimension is that of planning under uncertainty with *extended goals*. Intuitively, extended goals are conditions on the entire execution path of a plan. In several applications, indeed, goals may involve temporal conditions, e.g., conditions to be maintained rather than reached. For instance, a goal for an air-conditioner is to maintain the temperature in a desired range. Moreover, a goal may need to specify some safety condition that should be guaranteed all along plan execution. For instance, a mobile robot should always avoid dangerous areas. Goals may specify requirements of different strength that take into account nondeterminism and possible failures. For instance, we might require that the system "tries" to reach a certain state, and if it does not manage to do so, it guarantees that some safe state is maintained. As an example, we can require that a mobile robot tries to reach a given location, but guarantees to avoid dangerous rooms. *Planning under uncertainty with extended goals* is a challenging problem, since extended goals add a further complexity to the already complicated problem.

2 The Approach: Planning as Model Checking

Planning as Model Checking (see, e.g., Cimatti et al. (1997, 1998b,a); Giunchiglia and Traverso (1999); Daniele et al. (1999); Bertoli et al. (2001); Pistore and Traverso (2001); Cimatti et al.; Lago et al. (2002)) is an approach to planning under uncertainty that deals with nondeterminism, partial observability, and extended goals. Its key idea is to solve planning problems "model-theoretically". It is based on the following ingredients:

- A planning domain is a nondeterministic state transition system, where the same action may lead from the same state to many different states. The planner does not know which of the outcomes will actually take place when the action will be executed.

- Formulas in temporal logic express reachability goals, i.e. set of final desired states, as well as extended goals, i.e., conditions on the entire plan execution paths.

- Plans result in conditional and iterative behaviours. Depending on the source of uncertainty, plans encode conditional behaviours by mapping states into actions to be executed - in the case of full observability, they can express conditions over the result of observations - in the case of partial observability, or they may define the action to be executed also depending on the previous execution steps - in the case of extended goals.

- Given a state transition system and a temporal formula, planning by model checking generates plans that "control" the evolution of the system so that all of the system's behaviors make the temporal formula true. Plan validation can be formulated as a model checking problem Clarke et al. (1999).

- Planning algorithms use symbolic model-checking techniques Burch et al. (1992). In particular, sets of states are represented as propositional formulas, and searching through the state space is performed by doing logical transformations over propositional formulas. The algorithms are implemented by using Ordered Binary Decision Diagrams - OBDDs Bryant (1991), which allow for the compact representation and effective manipulation of propositional formulas.

In this paper, we describe first how the approach can deal with nondeterministic models under the assumption of simple "reachability goals" and full observability (Section 3). We then discuss how the approach can be extended to partial observability (Section 4), and finally to extended goals (Section 5).

3 Planning with Nondetermistic Models

A *planning domain* is a nondeterministic state-transition system $\Sigma = (S, A, \gamma)$ where S is a finite set of states, A is a finite set of actions, $\gamma : S \times A \rightarrow 2^S$ is the state-transition function. Non determinism is modeled by γ: given a state s and an action a, $\gamma(s, a)$ is a set of states. We say that an *action a is applicable in a state s* if $\gamma(s, a)$ is not empty. The set of actions that are applicable in state s is $A(s) = \{a : \exists s' \in \gamma(s, a)\}$.

Plans for nondeterministic domains are policies that encode conditional and iterative behaviours. A *policy* π for a planning domain $\Sigma = (S, A, \gamma)$ is a set of pairs (s, a) such that $s \in S$ and $a \in A(s)$. A *policy π is deterministic* if for any state s there is at most one action a such that $(s, a) \in \pi$; otherwise the policy is *nondeterministic*. The set of states of a policy is $S_\pi = \{s \mid (s, a) \in \pi\}$. A controller uses a reactive loop to execute a policy (see Figure 1). We represent the execution of a policy in a planning domain with

Execute-Policy(π)
    observe the current state s
    while $s \in S_\pi$ do
        choose an action a such that $(s, a) \in \pi$
        execute action a
        sense the current state s
end

Figure 1. The reactive loop for policy execution

an *execution structure*, i.e., a directed graph in which the nodes are all of the states of the domain that can be reached by executing actions in the policy, and the arcs represent a possible state transition caused by an action in the policy. Let π be a policy of a planning domain $\Sigma = (S, A, \gamma)$. The *execution structure* induced by π from the set of initial states $S_0 \subseteq S$ is a pair $\Sigma_\pi = (Q, T)$, such that $Q \subseteq S$, $T \subseteq S \times S$, $S_0 \subseteq Q$, and for every $s \in Q$ if there exists an action a such that $(s, a) \in \pi$, then for all $s' \in \gamma(s, a)$, $s' \in Q$ and $T(s, s')$. A state $s \in Q$ is a *terminal state* of Σ_π if there is no $s' \in Q$ such that $T(s, s')$.

Let $\Sigma_\pi = (Q, T)$ be the execution structure induced by a policy π from S_0. An *execution path* of Σ_π from $s_0 \in S_0$ is a possibly infinite sequence $s_0, s_1, s_2, \ldots$ of states in Q such that, for every state s_i in the sequence, either s_i is the last state of the sequence, in which case s_i is a terminal state of Σ_π, or $T(s_i, s_{i+1})$ holds. We say that a state s' is *reachable from* a state s if there is a path from s to s'.

A reachability goal is similar to a goal in classical planning, in that a plan succeeds if it reaches a state that satisfies the goal. However, since the execution of a plan may produce more than one possible path, the definition of a solution to a planning problem is more complicated than in classical planning. We distinguish among three possibilities:

1. *Weak solutions* are plans that may achieve the goal, but are not guaranteed to do so. A plan is a weak solution if there is at least one finite path that reaches the goal.
2. *Strong solutions* are plans that are guaranteed to achieve the goal in spite of non-determinism: all the paths are finite and reach the goal.
3. *Strong cyclic solutions* are guaranteed to reach the goal under a "fairness" assumption, i.e., the assumption that execution will eventually exit the loop. They are such that all their partial execution paths can be extended to a finite execution path whose terminal state is a goal state.

Weak and strong solutions correspond to the two extreme requirements for satisfying reachability goals. Intuitively, weak solutions correspond to "optimistic plans". Strong solutions correspond to "safe plans". However, there might be cases in which weak solutions are not acceptable, and strong solutions do not exist. In such cases, strong cyclic solutions may be a viable alternative.

A *planning problem* is a triple $\langle \Sigma, S_0, S_g \rangle$, where $\Sigma = \langle S, A, \gamma \rangle$ is a planning domain, $S_0 \subseteq S$ is a set of initial states, and $S_g \subseteq S$ is a set of goal states. Let π be a deterministic policy for Σ (see page 3). Let $\Sigma_\pi = \langle Q, T \rangle$ be the execution structure induced by π from S_0. Then:

1. π is a *weak solution* to P iff for any state in S_0, there exists a state in S_g that is a terminal state of Σ_π.
2. π is a *strong solution* to P iff Σ_π is acyclic and all of the terminal states of Σ_π are in S_g.
3. π is a *strong cyclic solution* to P iff from any state in Q there exists a terminal state of Σ_π that is reachable and all the terminal states of Σ_π are in S_g.

The set of strong solutions to a planning problem is a subset of the set of strong cyclic solutions, which in turn is a subset of the set of weak solutions.

A *determinization of* π is any deterministic policy $\pi_d \subseteq \pi$ such that $S_{\pi_d} = S_\pi$. From this it follows that π is a strong (or weak or strong cyclic) solution to a planning problem if all the determinizations of π are strong (or weak or strong cyclic) solution.

We can devise a planning algorithm for strong solutions that, given a planning problem $P = \langle \Sigma, S_0, S_g \rangle$ as input, either returns a policy that is a strong solution, or failure if a strong solution does not exists. The algorithm can be based on a breadth-first search proceeding backwards from the goal states toward the initial states. It iteratively computes a "preimage" of the current set of states, i.e., a current partial plan represented by the set of pairs state - action such that the action is guaranteed to lead to states in current set of states. The algorithm terminates if the initial states are included in the set of accumulated states, or if a fixed point has been reached from which no more states can be added to the current policy. In the first case, the returned policy is a solution to the planning problem. In the second case, no solution exists: indeed, there is some initial state from which the problem is not solvable. This algorithm is guaranteed to terminate. It is sound, i.e., the returned policies are strong solutions, and is complete, i.e., if it returns failure then there exists no strong solution. Moreover, it returns policies that are optimal in the following sense. A policy results in a set of paths. Consider the longest path of the policy, and let us call it *worst path*. Then the solution returned by the algorithm has a minimal *worst path* among all possible solutions. For the formal definition and proofs see Cimatti et al..

A planning algorithm for weak solutions is identical to the algorithm for strong planning, except that it iteratively computes a "weak preimage", i.e., a partial plan that may lead to the current set of states, rather than being guaranteed to do so.

The algorithm for strong cyclic planning computes a "greatest fixed point". It starts with the universal policy $\{(s, a) \mid a \in A(s)\}$ that contains all state-action pairs. It iteratively eliminates state-action pairs from the universal policy. We call the result at each step, the "current policy". This "elimination" phase, where state-action pairs leading to states out of the states of the current policy are discarded, is based on the repeated application of procedures that remove every state-action pair that leads out of the current set of states. Because of this elimination, from certain states it may become impossible to reach the set of goal states. Such states are identified and removed at each iteration. Due to this removal, the need may arise to eliminate further outgoing transitions, and so on. The elimination loop terminates when convergence is reached. The algorithm then checks whether the computed policy tells what to do in every initial state. If this is not the case, then a failure is returned. The algorithm is guaranteed to terminate, it is sound, i.e., the returned policies are strong cyclic solutions, and it is complete, i.e., if it returns failure then there exists no strong cyclic solution.

4 Planning under Partial Observability

In this section we address the problem of planning under partial observability for strong solutions, i.e. solutions that are guaranteed to reach a given set of states. A partially observable planning domain is a nondeterministic state transition system $\Sigma = (S, A, \gamma)$, where S, A, and γ are the same as in the previous section, a finite set O of observation variables with observations $\eta_o : S \times A \to \{true, false, unknown\}$, for any $a \in A$, $s \in S$, and $o \in O$. The function $\eta_o(s, a)$ gives the value of the observation variable o when the system is in state s and after action a has been executed. The value of o can be available to the controller, in which case either $\eta_o(s, a) = true$ or $\eta_o(s, a) = false$, and we say that o is *defined* in s after a. An observation may convey no information, in which case $\eta_o(s, a) = unknown$, and we say that o is *undefined* in s after a. The function $\eta_o(s, a)$ models partial observability, in which different states may convey the same observations. Let s and s' be two distinct states of Σ, i.e., $s, s' \in S$ and $s \neq s'$. We say that s and s' are *indistinguishable* if $\forall o \in O \ \forall a \in A \ . \ \eta_o(s, a) = \eta_o(s', a)$. The same state may convey different observations depending on the action that has been executed, i.e., we can have $\eta_o(s, a) \neq \eta_o(s, a')$. Intuitively, this says that the observations do not depend only on the state in which the system is, but also on the action that leads to that state. For instance, we can have some "sensing action" a that does not change the state, i.e., $\gamma(s,a) = \{s\}$, but that makes some observation available, e.g., for some $o \in O$, $\eta_o(s,a) \neq unknown$, while $\eta_o(s,a') = unknown$ for any $a' \neq a$. We say that an observation variable o is *action-independent* if it only depends on the state rather than on the action, i.e., $\forall a \in A \ \forall a' \in A \ \forall s \in S. \ \eta_o(s,a) = \eta_o(s,a')$.

In partially observable domains, the state of the system is normally not known. After an action is executed, the controller knows that the system is in a *set* of states, rather than exactly in *one* state. Sets of states are called *belief states*. Given a belief state, after the controller executes an action a, the controller can observe a new belief state b' that is obtained by applying the state transition function γ to all the states in b. We extend γ to work on belief states. Let b be a belief state, i.e., $b \subseteq S$. The belief state obtained by applying action γ to b is:

$$\gamma(b,a) = \{s' \mid s' \in \gamma(s,a) \text{ and } s \in b\} \tag{4.1}$$

We say that the action a is *applicable* in a non-empty belief state b iff a is applicable to all of the states of b, i.e., $\forall s \in b \ \exists s' \in S$, $s' \in \gamma(a,s)$. We call $A(b) = \{a \in A \mid a$ is applicable in $b\}$ the set of actions that is applicable in b. Notice that $\gamma(b,a) \neq \emptyset$ does not imply that a is applicable in b.

Given a belief state, an observation o partitions the belief state in the subsets of states where o is true, those where o is false, and those where o is unknown. We define $\eta_o^a(true) = \{s \in S \mid \eta_o(s,a) = true\}$, $\eta_o^a(false) = \{s \in S \mid \eta_o(s,a) = false\}$, and $\eta_o^a(unknown) = \{s \in S \mid \eta_o(s,a) = unknown\}$. If observations do not depend on actions, we write $\eta_o(true) = \{s \in S \mid \eta_o(s) = true\}$, $\eta_o(false) = \{s \in S \mid \eta_o(s) = false\}$, and $\eta_o(unknown) = \{s \in S \mid \eta_o(s) = unknown\}$. The state space S is partitioned in three partitions $\eta_o(true)$, $\eta_o(false)$, and $\eta_o(unknown)$, i.e., $\eta_o(true) \cup \eta_o(false) \cup \eta_o(unknown) = S$, and $\eta_o(true) \cap \eta_o(false) = \emptyset$, $\eta_o(true) \cap \eta_o(unknown) = \emptyset$, and $\eta_o(false) \cap \eta_o(unknown) = \emptyset$.

In partially observable domains, plans need to branch on conditions on the value of observable variables. We define a conditional plan:

- λ is a plan;
- If π_1 and π_2 are plans, then $\pi_1; \pi_2$ is a plan.
- If π_1 and π_2 are plans, and $o \in O$, then **if** o **then** π_1 **else** π_2 is a plan.

λ is the "empty plan", i.e., the plan that does nothing. Conditional plans need to "work on belief states" rather than on states. Given a belief state, they return a belief state. We therefore define $\Gamma(b,\pi)$, i.e., the execution of a plan π in a non empty belief state b (we restrict to the case of action-independent observation variables):

- $\Gamma(\emptyset, \pi) = \emptyset$; $\Gamma(b, \lambda) = b$;
- $\Gamma(b,a) = \gamma(b,a)$, if a is applicable in b,
  $\Gamma(b,a) = \emptyset$, otherwise;
- $\Gamma(b, \pi_1; \pi_2) = \Gamma(\Gamma(b, \pi_1), \pi_2)$;
- $\Gamma(b, \textbf{if } o \textbf{ then } \pi_1 \textbf{ else } \pi_2) = \Gamma(b \cap \eta_o(true), \pi_1) \cup \Gamma(b \cap \eta_o(false), \pi_2)$,
  if condition $app(b, \eta_0, \pi_1, \pi_2)$ holds
  $\Gamma(b, \textbf{if } o \textbf{ then } \pi_1 \textbf{ else } \pi_2) = \emptyset$, otherwise

where

$$app(b, \eta_0, \pi_1, \pi_2) = ((b \cap \eta_o(true) \neq \emptyset) \rightarrow \Gamma(b \cap \eta_o(true), \pi_1) \neq \emptyset) \land$$
$$(b \cap \eta_o(false) \neq \emptyset) \rightarrow \Gamma(b \cap \eta_o(false), \pi_2) \neq \emptyset$$

is a condition that states that if b has states where o is true, then the plan in the then-branch, i.e., π_1, must be applicable in these states, and, similarly, that if b has states where o is false, then the plan in the else-branch, i.e., π_2, must be applicable in these states. We do not need conditions in the states where o is unknown, since $\Gamma(b \cap \eta_o(unknown), \text{if } o \text{ then } \pi_1 \text{ else } \pi_2) = \emptyset$.

We say that a plan π is *applicable* in $b \neq \emptyset$ iff $\Gamma(b, \pi) \neq \emptyset$. If the plan is applicable, then its execution is the set of all states that can be reached after the execution of the plan.

A *planning problem* is the tuple $\langle \Sigma, b_0, b_g \rangle$, where Σ is a planning domain, $b_0 \subseteq S$ is the initial belief state, and $b_g \subseteq S$ is the goal belief state. We require that b_0 and b_g are not empty. A plan π is a *strong solution* to the problem $\langle \Sigma, b_0, b_g \rangle$ iff π is applicable to the initial belief state, i.e., $\Gamma(b_0, \pi) \neq \emptyset$, and the execution of π in the initial belief state results in a belief state that is a subset of the goal belief state, i.e., $\Gamma(b_0, \pi) \subseteq b_g$.

An algorithm for planning under partial observability can search an and-or graph over belief states. The and-or graph can be recursively constructed from the initial belief state b_0, expanding each encountered belief state by every possible combination of applicable actions and observations.

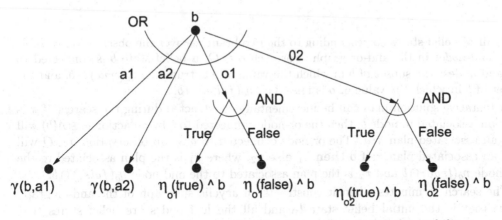

Figure 2. The search space for planning under partial observability

Figure 2 shows an example of an and-or graph on belief states. Each node in the graph is a belief state. A node b can be connected to or-nodes or to and-nodes. The *or-nodes* connected to a node b are

- either the belief states produced by the applicable actions $a \in A(b)$. For each $a \in A(b)$, we have an or-node $\Gamma(b, a)$ connected to node b,
- or the pairs of belief states produced by observations $o \in O$. Each pair corresponds to two branches, one in which o is true and the other in which o is false.

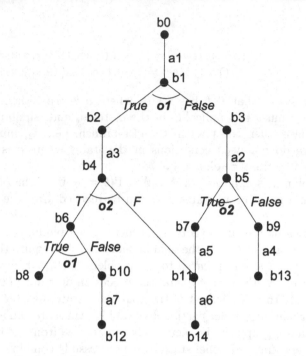

Figure 3. An and-or graph corresponding to a conditional plan

The pair of belief states corresponding to the two branches over the observation variable o are *and-nodes* in the and-or graph. For each $o \in O$, a belief state b is connected to two and-nodes: the subset of b in which the value of o is true, i.e., $\eta_o(true) \cap b$, and the subset of b in which the value of o is false, i.e., $\eta_o(false) \cap b$.

A tentative solution plan can be incrementally constructed during the search. If π is the plan associated to node b, then the or-node connected to b by an action $a \in A(b)$ will have an associated plan $\pi; a$. The or-node connected to b by an observation $o \in O$ will have an associated plan $\pi; \text{if } o \text{ then } \pi_1 \text{ else } \pi_2$, where π_1 is the plan associated to the and-node $\eta_o(true) \cap b$ and π_2 is the plan associated to the and-node $\eta_o(false) \cap b$.

The search terminates when it constructs an acyclic subgraph of the and-or graph whose root is the initial belief state b_0 and all the leaf nodes are belief states that are subsets of the goal belief state b_g. Figure 3 shows a subgraph that corresponds to the plan $a1; \text{if } o1 \text{ then } \pi_1 \text{ else } \pi_2$, where $\pi_1 = a3; \text{if } o2 \text{ then } \pi_3 \text{ else } \pi_4$, and $\pi_2 = a2; \text{if } o2 \text{ then } \pi_5 \text{ else } \pi_6$, and so on.

A planning algorithm can search this and-or graph in different ways and with different strategies. For instance, a breadth first forward search, given a belief state b proceeds by exploring all the possible or-nodes. The termination condition is reached when we have at least one and-or subgraph that meets the conditions described above. Breadth first search can generate optimal plans but it can be very expensive. A depth first search selects one of the possible or-nodes and needs to explore each branch of the selected and-nodes. It can be much more convenient than depth first search, but it does not guarantee to return optimal plans. Different heuristics can be applied in the depth first search, e.g., select all the observations first, select the or-nodes that produce smaller belief states etc.

5 Planning for Extended Goals

In this section we address the problem of planning for goals expressed as formulas in temporal logic. We restrict to the case of full observability. A planning domain is a state transition system $\Sigma = (S, A, \gamma)$, where S, A, and γ are the same as in previous sections, except for one minor difference. In order to simplify the definitions, we require γ to be total, i.e., $\forall s\ A(s) \neq \emptyset$. We call L a set of basic propositions such that $S \subseteq 2^L$. Each basic proposition $b \in L$ may either hold in a give state s, and we write $b \in s$, or not hold, and we write $b \notin s$. A goal g for Σ is defined as follows:

$$g ::= \quad \top \mid \bot \mid b \mid \neg b \mid g \to g \mid g \land g \mid g \lor g \mid \mathrm{AX}\,g \mid \mathrm{EX}\,g \mid$$
$$\mathrm{A}(g\,\mathrm{U}\,g) \mid \mathrm{E}(g\,\mathrm{U}\,g) \mid \mathrm{A}(g\,\mathrm{W}\,g) \mid \mathrm{E}(g\,\mathrm{W}\,g)$$

Goals are essentially CTL formulas (Emerson (1990)) A remark is in order: even if $\neg$ is allowed only in front of basic propositions, it is easy to define $\neg g$ for a generic CTL formula g, by "pushing down" the negations: for instance $\neg \mathrm{AX}\,g \equiv \mathrm{EX}\,\neg g$ and $\neg\mathrm{A}(g_1\,\mathrm{W}\,g_2) \equiv \mathrm{E}(\neg g_2\,\mathrm{U}(\neg g_1 \land \neg g_2))$. CTL allows us to specify different kinds of goals. For instance, weak, strong, and strong cyclic solutions, can be expressed with the formulas $\mathrm{EF}\,g$, $\mathrm{AF}\,g$, and $\mathrm{A}(\mathrm{EF}\,g\,\mathrm{W}\,g)$, respectively. Intuitively $\mathrm{AG}\,g$ means "maintain g", $\mathrm{AG}\,\neg g$ means "avoid g", $\mathrm{EG}\,g$ means "try to maintain g", and $\mathrm{EG}\,\neg g$ means "try to avoid g". The CTL formula $\mathrm{AF}\,\mathrm{AG}\,g$ states that a plan should guarantee that all executions reach eventually a set of states where g can be maintained, while $\mathrm{AG}\,\mathrm{EF}\,g$ means "maintain the possibility of reaching g".

In the case of extended goals, plans as policies are not enough. Different actions may need to be executed in the same state depending on the previous states of the execution path. Plans must take into account the "context of execution", i.e. the internal state of the controller. A *plan* for a domain Σ is a tuple $(C, c_0, act, ctxt)$, where:

- C is a set of (execution) contexts,
- $c_0 \in C$ is the initial context,
- $act: S \times C \rightharpoonup A$ is the action function,
- $ctxt: S \times C \times S \rightharpoonup C$ is the context function.

If we are in state s and in execution context c, then $act(s, c)$ returns the action to be executed by the plan, while $ctxt(s, c, s')$ associates to each reached state s' the new execution context. Functions act and $ctxt$ are partial, since some state-context pairs are never reached in the execution of the plan.

We say that plan π is *executable* if, whenever $act(s, c) = a$ and $ctxt(s, c, s') = c'$, then $s' \in \gamma(s, a)$. We say that π is *complete* if, whenever $act(s, c) = a$ and $s' \in \gamma(s, a)$, then there is some context c' such that $ctxt(s, c, s') = c'$ and $act(s', c')$ is defined. Intuitively, a complete plan always specifies how to proceed for all the possible outcomes of any action in the plan. In the following, we consider only plans that are *executable and complete*.

The execution of a plan results in a change in the current state and in the current context. It can therefore be described in terms of transitions from one state-context pair to another. Formally, given a domain Σ and a plan π, a transition of plan π in Σ is a tuple $(s,c) \stackrel{a}{\rightarrow} (s',c')$ such that $s' \in \gamma(s,a)$, $a = act(s,c)$, and $c' = ctxt(s,c,s')$. A *run* of plan π from state s_0 is an infinite sequence $(s_0, c_0) \stackrel{a_0}{\rightarrow} (s_1, c_1) \stackrel{a_1}{\rightarrow} (s_2, c_2) \stackrel{a_2}{\rightarrow} (s_3, c_3) \cdots$ where $(s_i, c_i) \stackrel{a_i}{\rightarrow} (s_{i+1}, c_{i+1})$ are transitions.

The *execution structure* of plan π in a domain Σ from state s_0 is the structure $\Sigma_\pi = (Q, T, L)$, where:

- $Q = \{(s,c) \mid act(s,c) \text{ is defined}\}$,
- $((s,c), (s',c')) \in T$ if $(s,c) \stackrel{a}{\rightarrow} (s',c')$ for some a,
- $L(s,c) = \{b \mid b \in s\}$

A *planning problem* is the tuple (Σ, S_0, g), where Σ is a nondeterministic state transition system, $S_0 \subseteq S$ is a set of initial states, and g is a goal for Σ. Let π be a plan for Σ and Σ_π be the corresponding execution structure. A plan π satisfies goal g from initial state $s_0 \in S$, written $\pi, s_0 \models g$, if $\Sigma_\pi, (s_0, c_0) \models g$. A plan π satisfies goal g from the set of initial states S_0 if $\pi, s_0 \models g$ for each $s_0 \in S_0$.

Σ_π is a Kripke Structure (Emerson (1990)). (s_0, c_0) is the initial state of the Kripke Structure. In model checking, Σ_π is the model of a system and g is a property to be verified. The plan validation problem, i.e., the problem of determining whether a plan satisfies a goal, is thus formulated as the model checking problem of determining whether the CTL formula g is true in the Kripke Structure Σ_π, which represents the behaviours of the system Σ "controlled" by the plan π. The plan generation problem can be solved with algorithms that progress the goal formula, see, e.g., Pistore and Traverso (2001); Lago et al. (2002).

6 Related Work and Conclusions

Planning as Model Checking is a promising approach to tackle the problem of generating plans with nondeterministic models, under partial observability, and for extended goals. There are still many problems to be solved, among them, the problem of planning with preferences. Furthermore, the problem of plan generation under uncertainty is rather similar to the problem of the automatic synthesis of reactive controllers. The potentiality of the planning as model checking approach in this field has still to be explored. Moreover, in many applications, the plan generation problem should not be separated by the execution problem, since the idea of interleaving planning and execution seems a very reasonable one to tackle uncertainty at the loss of completeness. Finally, it is still to be investigated if the approach can be used to do real reactive planning.

The planning as model checking approach has been exploited for planning under partial observability (Rintanen (2002)), and in the work on strong, strong cyclic and optimistic planning described in Jensen and Veloso (2000). Jensen et al. (2001) addresses the problem of using model checking to do adversarial planning with contingent events. Model checking techniques are also used for planning for LTL goals in Bacchus and Kabanza (1996b) and for planning under hard real time constraints in Goldman et al. (2000, 1999, 1997). De Giacomo and Vardi (1999) propose an automata theoretic approach. Finally, BDD-based symbolic model checking has also been used to tackle the problem of planning in deterministic domains, see Edelkamp and Helmert (2000, 1999), and to do hand-tailored planning in TL-plan, see Bacchus and Kabanza (1996a).

Bibliography

F. Bacchus and F. Kabanza. Using temporal logic to control search in a forward chaining planner. In M. Ghallab and A. Milani, editors, *New Directions in AI Planning*, pages 141–156. IOS Press (Amsterdam), 1996a. ISBN 90-5199-207-0.

Fahiem Bacchus and Froduald Kabanza. Planning for temporally extended goals. In *Proceedings of the Thirteenth National Conference on Artificial Intelligence and the Eighth Innovative Applications of Artificial Intelligence Conference*, pages 1215–1222, Menlo Park, August4–8 1996b. AAAI Press / MIT Press. ISBN 0-262-51091-X.

P. Bertoli, A. Cimatti, M. Roveri, and P. Traverso. Planning in nondeterministic domains under partial observability via symbolic model checking. In B. Nebel, editor, *Proceedings of the Seventeenth International Joint Conference on Artificial Intelligence, IJCAI 2001*, pages 473–478. Morgan Kaufmann Publishers, August 2001.

R. E. Bryant. On the complexity of VLSI implementations and graph representations of Boolean functions with application to integer multiplication. *IEEE Transactions on Computers*, 40(2):205–213, February 1991.

J. R. Burch, E. M. Clarke, K. L. McMillan, D. L. Dill, and L. J. Hwang. Symbolic Model Checking: 10^{20} States and Beyond. *Information and Computation*, 98(2):142–170, June 1992.

A. Cimatti, E. Giunchiglia, F. Giunchiglia, and P. Traverso. Planning via Model Checking: A Decision Procedure for AR. In S. Steel and R. Alami, editors, *Proceeding of the Fourth European Conference on Planning*, number 1348 in Lecture Notes in Artificial Intelligence, pages 130–142, Toulouse, France, September 1997. Springer-Verlag.

A. Cimatti, M. Pistore, M. Roveri, and P. Traverso. Weak, Strong, and Strong Cyclic Planning via Symbolic Model Checking. *Artificial Intelligence*.

A. Cimatti, M. Roveri, and P. Traverso. Automatic OBDD-based Generation of Universal Plans in Non-Deterministic Domains. In *Proceeding of the Fifteenth National Conference on Artificial Intelligence (AAAI-98)*, Madison, Wisconsin, 1998a. AAAI-Press. Also IRST-Technical Report 9801-10, Trento, Italy.

A. Cimatti, M. Roveri, and P. Traverso. Strong Planning in Non-Deterministic Domains via Model Checking. In *Proceeding of the Fourth International Conference on Artificial Intelligence Planning Systems (AIPS-98)*, Carnegie Mellon University, Pittsburgh, USA, June 1998b. AAAI-Press.

E. M. Clarke, O. Grumberg, and D. A. Peled. *Model Checking*. The MIT Press, Cambridge, Massachusetts, 1999. ISBN 0262032708.

M. Daniele, P. Traverso, and M. Y. Vardi. Strong cyclic planning revisited. In S. Biundo, editor, *Proceeding of the Fifth European Conference on Planning*, volume 1809 of *LNAI*, Durham, United Kingdom, September 1999. Springer-Verlag.

G. De Giacomo and M. Y. Vardi. Automata-theoretic approach to planning for temporally extended goals. In S. Biundo, editor, *Proceeding of the Fifth European Conference on Planning*, volume 1809 of *LNAI*, pages 226–238, Durham, United Kingdom, September 1999. Springer-Verlag.

S. Edelkamp and M. Helmert. Exhibiting knowledge in planning problems to minimize state encoding length. In S. Biundo and M. Fox, editors, *Proceedings of the Fifth*

European Conference on Planning (ECP'99), volume 1809 of *LNAI*, pages 135–147. Springer-Verlag, 1999.

S. Edelkamp and M. Helmert. On the implementation of mips. In *AIPS-Workshop on Model-Theoretic Approaches to Planning*, pages 18–25, 2000.

E. A. Emerson. Temporal and modal logic. In J. van Leeuwen, editor, *Handbook of Theoretical Computer Science, Volume B: Formal Models and Semantics*, chapter 16, pages 995–1072. Elsevier, 1990.

F. Giunchiglia and P. Traverso. Planning as model checking. In S. Biundo, editor, *Proceeding of the Fifth European Conference on Planning*, volume 1809 of *LNAI*, pages 1–20, Durham, United Kingdom, September 1999. Springer-Verlag.

R. P. Goldman, D. J. Musliner, K. D. Krebsbach, and M. S. Boddy. Dynamic abstraction planning. In *Proceedings of the Fourteenth National Conference on Artificial Intelligence and Ninth Innovative Applications of Artificial Intelligence Conference (AAAI 97), (IAAI 97)*, pages 680–686. AAAI Press, 1997.

R. P. Goldman, D. J. Musliner, and M. J. Pelican. Using Model Checking to Plan Hard Real-Time Controllers. In *Proceeding of the AIPS2k Workshop on Model-Theoretic Approaches to Planning*, Breckeridge, Colorado, April 2000.

R.P. Goldman, M. Pelican, and D.J. Musliner. Hard Real-time Mode Logic Synthesis for Hybrid Control: A CIRCA-based approach, mar 1999. Working notes of the 1999 AAAI Spring Symposium on Hybrid Control.

R. Jensen and M. Veloso. OBDD-based Universal Planning for Synchronized Agents in Non-Deterministic Domains. *Journal of Artificial Intelligence Research*, 13:189–226, 2000.

R. M. Jensen, M. M. Veloso, and M. H. Bowling. OBDD-based optimistic and strong cyclic adversarial planning. In *Proceedings of the Sixth European Conference on Planning (ECP'01)*, 2001.

U. Dal Lago, M. Pistore, and P. Traverso. Planning with a language for extended goals. In *In Proceedings of the Eighteenth National Conference on Artificial Intelligence (AAAI-02)*, 2002.

M. Pistore and P. Traverso. Planning as model checking for extended goals in non-deterministic domains. In B. Nebel, editor, *Proceedings of the Seventh International Joint Conference on Artificial Intelligence (IJCAI-01)*, pages 479–486. Morgan Kaufmann Publisher, August 2001.

J. Rintanen. Backward plan construction for planning as search in belief space. In *In Proceedings of the Sixth International Conference on AI Planning and Scheduling (AIPS'02)*, 2002.

Understanding Control Strategies

Ivan Bratko and Dorian Šuc

Faculty of Computer and Information Sc., University of Ljubljana, Ljubljana, Slovenia

Abstract. Controlling complex dynamic systems requires skills that operators often cannot completely describe, but can demonstrate. This paper describes research into the understanding of such tacit control skills. Understanding tacit skills has practical motivation in respect of communicating skill to other operators, operator training, and also mechanising and optimising human skill. This paper is concerned with approaches to the understanding of human operators' skill by analysing operators' traces using techniques of machine learning (ML). The paper gives a review of ML-based approaches to skill reconstruction. Recent work is presented with particular emphasis on understanding human tacit skill *qualitatively*, and automatically generating explanation of how it works. This includes qualitative machine learning to extract from operator's control traces his subconscious sub-goals and control strategies described in qualitative terms.

1 Introduction

Controlling a complex dynamic system, such as an aircraft or a crane, requires operator's skill acquired through experience. In this paper we are interested in the question of understanding tacit control skills which can be the basis for designing automatic controllers that employ the sub-cognitive control principles used by the operator. One approach would be to attempt to extract the skill from the operator in a dialogue fashion whereby the operator would be expected to describe his or her skill. This description would then be appropriately formalised and built into an automatic controller.

The problem with this approach is that the skill is sub-cognitive and the operator is usually only capable of describing it incompletely and approximately. Such descriptions can only be used as vague and unreliable guidelines for constructing automatic controllers. As discussed for example in (Urbančič and Bratko, 1994; Bratko and Urbančič 1999), the operator's descriptions are not operational in the sense of being directly translatable into an automatic controller.

Given the difficulties of skill transfer through introspection, an alternative approach to skill reconstruction is to start from the *manifestation* of the skill. Although an operational description of the skill is not available, the manifestation of the skill *is* available in the form of traces of the operator's actions. One idea is to use these traces as examples and extract operational models of the skill by Machine Learning (ML) techniques. This has been explored in a number of experiments. Sammut et al. (1992) present one of the earlier and best-known such studies.

Extracting models of a real-time skill from operators' behaviour traces by Machine Learning is also known as *behavioural cloning* (Michie 1993; Michie et al. 1990). In general there are two goals of behavioural cloning:

- To generate *good performance* clones, that is clones that can reliably carry out the control task.
- To generate *meaningful* clones, that is clones that would help making the human operator's skill symbolically explicit.

The second goal is important for several reasons:

- Operationalising human operator's instructions for controlling a system. Such instructions are a useful source of information, but are normally too incomplete and imprecise to be directly translatable into a control program.
- Flexibly modifying and optimising the induced clones to prevent some undesired patterns in their behaviour.
- Understanding what exactly a human operator is doing and why. This is of practical importance regarding the capture of exceptional operators' skill and its transfer to less gifted operators. The operator's control strategy would ideally be understood in terms of goals, sub-goals, plans, feedback loops, causal relations between actions and state conditions etc. These conditions are to be stated in terms of information that is easily accessible to the operator, e.g. visually. It can be argued that such information should be largely qualitative, as opposed to the prevailing numerical information.

Behavioural cloning has been studied in various dynamic domains, including: pole balancing (Michie et al. 1990), flying a Cessna aircraft (Sammut et al. 1992; Bain and Sammut 1999), flying the F16 (Michie and Camacho, 1994), operating crains (Urbančič and Bratko, 1994; Šuc and Bratko 1997), electrical discharge machining (Karalič and Bratko 1997), the "acrobot" (double linked pendulum; Šuc and Bratko 2000b). Figure 1 illustrates two of these dynamic systems.

2 Inducing "direct controllers"

The following is the usual procedure of applying Machine Learning (ML) to recovering control strategy from example execution traces. A continuous trace is sampled so that we have a sequence of pairs ($State_i$, $Action_i$) ordered according to time. $State_i$ is the state of the system at time i, and $Action_i$ is the operator's action performed at time i. Then, usually, the sequence of these pairs is viewed as a *set* of examples, thereby ignoring the time order. This simplification can be justified by the formal argument that the control action is fully determined by the state of the controlled system. Many machine learning programs can be viewed as reconstructors of functions $y = f(x_1, x_2, ...)$ from sets of given pairs of the form:

$((x_{1i}, x_{2i}, ...), y_i)$

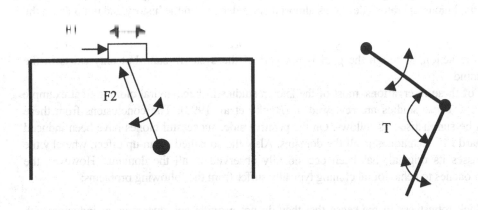

Figure 1. Two frequently used dynamic systems in experiments in behavioural cloning. The arrows indicate degrees of freedom and possible control actions. Left: container crane with horizontal (F1) and vertical (F2) control forces; the task is to move the load to a goal position while controlling the swinging of the rope at the goal. Right: the "acrobot" (double-linked pendulum) with control torque T applied at the middle joint; the task is to gradually increase the oscillation of the links until the linkage makes a full cycle over the top bar.

The arguments x_1, x_2, ... are usually called attributes, and function value y is called the *class* value. So the usual formulation of behavioural cloning as a ML task is as follows: the system's state variables x_1, x_2 etc. are considered as attributes, and the action as the class variable. In behavioural cloning, the most frequently used ML techniques have been decision tree learning (Quinlan 1986) and its variation regression tree learning (Breiman et. al. 1984; Karalic 1992). In fact, any technique that reconstructs a function from examples can be used, including neural networks. However, most studies in behavioural cloning are limited to the use of symbolic learning techniques, thus excluding neural networks. Neural networks would not be appropriate with respect to the goal of understanding the operator's skill. In addition to attaining good performance clones, we are also interested in understanding the operator's control strategy. What is the operator's control strategy? Why is he doing this or that?

The result of machine learning, using the formulation of the learning problem above, is a controller in the form of a function from system states to actions:

$$Action = f(State) = f((x_1, x_2, ...))$$

This controller maps the system's current state into an action *directly*, without any intermediate, auxiliary result. Therefore such controllers will be called *direct controllers*, to be distinguished from "indirect" controllers discussed later.

It has been shown empirically that direct controllers, although they appear to be the most natural, suffer from serious limitations. Part of the reason of direct controllers' inferiority can be explained by the following. The representational decision of viewing sequences as sets is debatable: the human operator's decisions almost always depend on the history and not only on the current state of the system. In controlling the system, the operator pursues certain goals over a time period and then switches to other goals, etc. Therefore, both the state *and* the current goal determine the action, although the goal is not part of the system's state but only exists in the operator's mind.

In spite of these reservations, most of the known studies in behavioural cloning treat example traces as sets. These studies are reviewed in (Bratko et al. 1998). The conclusions from these studies can be summarised as follows. On the positive side, successful clones have been induced using standard ML techniques in all the domains. Also, the so-called clean-up effect, whereby the clone surpasses its original, has been occasionally observed in all the domains. However, the present approaches to behavioural cloning typically suffer from the following problems:

- They lack robustness in the sense that they do not provide any guarantee of inducing with high probability a successful clone from given data.
- Typically, the induced clones are not sufficiently robust with respect to changes in the control task.
- Although the clones do provide some insight into the control strategy, they in general lack conceptual structure and representation that would clearly capture the structure and style of the operator's control strategy.

The reasons for mixed success of direct controllers usually lie in the representation used in skill reconstruction. The usual representation is inherited from traditional control theory, and is entirely numerical and unstructured. More appropriate representations are largely qualitative and involve goals and history (not just the current state of the system) and qualitative trends.

3. Inducing indirect controllers

We say that a controller is "indirect" if it does not compute the next action directly from the current system's state, but uses in addition to the state some other, intermediate information. Typical such additional information is a sub-goal to be attained before attaining the final goal.

Subgoals often feature in operator's control strategies. Their skill is also often best explained in terms of subgoals. The first idea to exploit this fact is to automatically detect, from operator's traces, such subgoals. This was studied in (Šuc and Bratko 1997) with an approach based on the theory of linear QR controllers. The problem of subgoal identification was treated as the inverse of the usual problem of controller design. The usual problem is: given a system's model and a goal, find the actions that achieve the goal. The problem of goal identification is: given the actions (in operator's trace), find the goal that these actions achieve. More precisely, find the goal so that the actions performed are QR-optimal with respect to this goal.

This approach was applied successfully to behavioural cloning in the crane domain (Šuc and Bratko 1997). However, the limitation of the approach is that it only works well for the cases in which there are just a few subgoals. A counter example is the acrobot domain. The task of completing a full cycle requires a sequence of oscillations, each of them having an increased amplitude. Such a trajectory consists of dense subgoals (illustrated in Figure 2). For such cases a more appropriate idea is to generalise the operator's trajectory. Such a generalised trajectory can be viewed as defining a continuously changing subgoal (Figure 2)

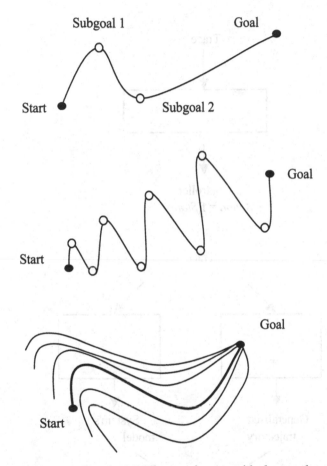

Fig. 2 Top: a trajectory and two subgoals. Middle: a trajectory with dense subgoals. Bottom: a "generalised trajectory".

Subgoals and generalised trajectories are not sufficient to define a controller. A model of the system's dynamics is also required. Therefore, in addition to inducing subgoals or a generalised trajectory, this approach also requires the learning of approximate system's dynamics, that is a model of the controlled system. The next action is then computed "indirectly", targeting the subgoals as follows: (1) compute the desired next state (e.g. next sub-goal), and (2) determine an action that brings the system to the desired next state. Alternatively, when trying to follow a generalised trajectory, the action is determined indirectly as: using the model of the system's dynamics, and the generalised trajectory, find the action that will minimise the difference between the generalised trajectory and the state resulting from this action.

Figure 3 illustrates this approach and contrasts it to the induction of direct controllers. The next action is computed so as to minimise the difference between the current state of the system and

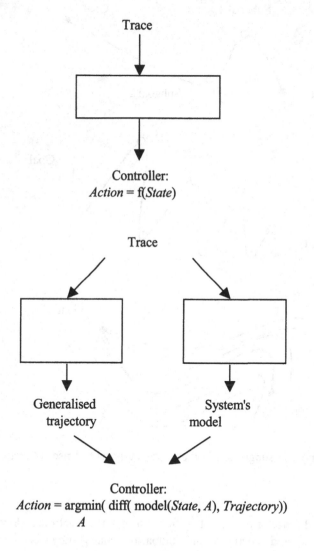

Fig. 3 Top: induction of direct controllers. Bottom: induction of indirect controllers.

the generalised trajectory. This can be done in various ways, and the "difference" between the current state and the trajectory is defined accordingly.

The point of indirect controllers is that the problem of behavioural cloning is decomposed into two learning problems, (1) learning trajectory, and (2) learning dynamics. It has been shown experimentally that this decomposition (Šuc and Bratko 2000a) leads to much better performance than the induction of direct controllers. The overall controller's performance of the induced clones is less sensitive to the classification accuracy achieved in the two learning tasks than in the case of direct controllers. This is demonstrated for the crane domain in (Šuc and Bratko 1999). In that work, the two learning tasks were accomplished by Križman's program Goldhorn (Križman et al. 1995) for inducing generalised trajectory, and locally weighted regression for learning approximate system's dynamics. Ideas of using subgoals in behavioural cloning in aeroplane flying has also been discussed by Bain and Sammut (1999).

4 Using qualitative representations

4.1 Numerical vs. qualitative representations

An important representational decision in behavioural cloning is concerned with the choice between a numerical representation, or a qualitative representation of induced descriptions. For example, whether to use system state variables as attributes for machine learning, or some other, qualitative attributes of the given traces.

An operator, executing his skill on an actual physical system, or on a graphical simulator, hardly uses the numerical state variables. First, some of these variables values are hard to observe precisely (as the classical control theory would assume), for example the angular velocity of the rope in crane control. In addition, the operator surely does not online evaluate an arithmetic formula to decide on his next action. Instead, he has to base his control decisions on other, easily recognisable patterns in terms of qualitative features of the current state and recent behaviour of the system. Such qualitative features include the following ones, well known in qualitative physics: the sign of a variable (for example, in crane control, the rope angle positive or negative), the direction of change (rope swinging left or right), variable crossing a landmark (rope upright), variable just reached a local extreme (load at an extreme swing position).

There is good experimental evidence that in behavioural cloning qualitative representations are more suitable than numerical representations. The use of qualitative attributes in the induction of direct controllers for the pole-and-cart task was studied by Bratko (1997). It was shown that the qualitative approach resulted in clones whose style of control was much closer to the original human operator's control. The resulting qualitative controllers also provided a much better explanation of the operator's strategy.

4.2 Monotonic functional constraints

One way of using qualitative representation in indirect controllers was studied by Šuc (2001). A generalised operator's trajectory can be described in terms of qualitative constraints typically used in qualitative physics such as monotonically increasing or decreasing functions. For example, the constraint $Y = M^+(X)$ says that Y is a monotonically increasing function of X. Analogously, $Y = M^-(X)$ says that Y is a monotonically decreasing function of X. These constraints can have multiple arguments. For example, $Z = M^{+,-}(X,Y)$ means that Z is monotonically increasing in X and decreasing in Y. If both X and Y increase then Z may increase, decrease, or stay unchanged.

We now illustrate how a control strategy can be described qualitatively, using such monotonicity constraints. For example, consider the crane at rest in the initial state. To start the crane moving, both velocity V and position X should be increasing simultaneously. This can be stated by the usual qualitative constraint:

$$V = M^+(X)$$

It should be noted that this is not a law of the physics of the crane system, but it is a *control* law enforced by a controller. Another qualitative rule about controlling the swing of the load may be: the greater the rope angle and the faster the angle is increasing, the greater should be the carriage velocity to make the carriage "catch up" with the angle. This can be stated by the constraints:

$$V = M^{+,+}(Fi, DFi)$$

4.3 Qualitative trees

Monotonicity constraints can be combined into if-then rules to express piece-wise monotonic functional relationships. For example:

$$\text{if } X < 0 \text{ then } Y = M^-(X) \text{ else } Y = M^+(X)$$

Nested if-then expressions can be represented as trees, called *qualitative trees* (Šuc 2001). Qualitative trees are similar to regression trees (Breiman et al. 1985). Both regression and qualitative trees describe how a numerical variable (called *class*) depends on other (possibly numerical) variables (called *attributes*). The difference between the two types of trees only occurs in the leaves. A leaf of a regression tree specifies a numerical regression function that tells how the class variable numerically depends on the attributes within the scope of the leaf. On the other hand, a leaf in a qualitative tree only specifies the relation between the class and the attributes *qualitatively*, in terms of monotonic qualitative constraints.

4.4 Learning of qualitative trees

QUIN (Qualitative Induction) is a learning program that induces qualitative trees from numerical data. QUIN detects monotonic qualitative constraints that hold "sufficiently well" in the data. QUIN employs a criterion of "qualitative fit" between a qualitative tree and the learning examples. This criterion takes into account the consistency of the tree with the examples, and the "ambiguity" of the tree. The tree is *inconsistent* with a pair of example points if it makes an incorrect qualitative prediction. For example, the class value between the two points increases, and the tree says that it should decrease. The tree is *ambiguous* with respect to a pair of example points if it permits, in addition to the correct qualitative change of the class between the two points, also changes in other directions. The size of a qualitative tree, and its degree of inconsistency and ambiguity, are combined into an encoding length cost. In the spirit of the MDL principle ⟨⟨⟨⟨⟨⟨⟨⟨⟨⟨⟨⟨ ⟨⟨⟨⟨⟨⟨⟨⟨⟨⟨ ⟨⟨⟨⟨⟨⟨⟨⟩ ⟨ ⟨⟨ ⟨⟨⟨ ⟨⟨⟨⟨⟨⟨⟨⟨ ⟨⟨⟨ ⟨⟨⟨⟨ ⟨⟨⟨⟨⟨⟨⟨⟨⟨⟨⟨⟨ ⟨⟨⟨⟨ ⟨⟨⟨⟨⟨⟨⟨⟨ ⟨⟨ ⟨⟨⟨ ⟨⟨⟨⟨⟨⟨⟨⟨ ⟨⟨⟨⟨ These mechanisms also make QUIN relatively robust in the case of noisy data. The details of this and the QUIN learning algorithm can be found elsewhere (Šuc 2001; Šuc and Bratko 2001).

5 Understanding of crane control strategies through qualitative analysis

In this section we look at an application of QUIN to qualitatively analyse control traces of human operators when controlling a simulated crane of the type shown in Figure 1 (Gantry crane). Each control trace is a sequence of dynamic states of the crane system and operator's actions performed in these states. A control trace in this domain typically consisted of 500 to 1000 states taken at sample time points. In the case of the Gantry crane, a dynamic state of the system consists of six variables:

X, position of the trolley
DX, trolley velocity
Fi, angle of rope with respect to vertical
DFi, angular velocity
L, length of rope
DL, rope length velocity

The possible actions are: force F1 (to push the trolley left or right), and force F2 (to pull the rope). In the task executed by the operator, in the initial state the trolley's position was X = 0. The trolley's goal position was X = 60. The requirement was that when the load was at the goal position, the swinging of the load was within very small tolerance around the vertical. The task was to be executed in as short a time as possible.

To qualitatively analyse a control trace with QUIN, we have to choose one of the state variables or an action variable as the class variable, and a subset of the remaining state variables as the attributes. Let us consider just the horizontal movement of the trolley and the swinging of the load, and choose DX as the class variable and X, Fi and DFi as the attributes. With this selection of variables, we use QUIN to find qualitative properties of the relation DX = f(X, Fi, DFi). It should be noted that this is just one possible choice, and that other choices of class and attributes would also possibly provide different useful characterisations of control strategies.

Below we give as an example two qualitative trees induced by QUIN from qualitative traces of two of our human operators that we here refer to S and L. We chose these two operators because their control styles were rather different and we were interested in finding the differences in their control strategies. Operator S performed very cautiously, never causing a large swing of the load. To avoid large swing, he could not afford large accelerations of the trolley. The qualitative tree (written as an if-then expression) induced from a trace of operator S, is:

if X < 20.7
    then DX = M⁺(X)
    else
    if X < 60.1 then DX = M⁻(X) else DX = M⁺(Fi)

This tree offers a nice explanation of operator S's control strategy. In the initial stage, when X is small, the desired velocity DX increases with X. From about one third of the total distance to the goal, velocity DX starts decreasing with X. Obviously, this aims at zero velocity when the goal X = 60 is reached. So far the qualitative tree corresponds to our prior belief about S's control strategy. We believed that this operator did not possess any skill of controlling the swing, and therefore he managed to complete the task by never causing a large swing by quick accelerations. Accordingly, the rope angle in S's control traces was always kept close to vertical. We believed that, to cope with small oscillations of the rope, this operator just waited until such an oscillation died out in time due to friction. Thus the right-most leaf of the qualitative tree above (DX = M⁺(Fi)) came as a surprise because it mentions the angle Fi. This reveals that this operator in fact did actually make an effort at controlling the swing, although only at the very final stage.

By just observing operator L, it was immediately obvious that this operator was much more adventurous than S. His skill allowed him to afford vigorous accelerations in the X direction, causing large swing of the load. However, as he had the skill of reducing the swing later, he was not afraid of large swing. This enabled him to perform much more efficiently in terms of execution time, achieving about 20 or 30% shorter finishing times than S. A qualitative tree induced by QUIN from a trace of operator L is as follows:

if X < 29.3
    then DX = M⁺·⁻·(X, Fi, DFi)
    else
    if DFi < -0.02 then DX = M⁻(X) else DX = M⁻·⁺(X, Fi)

Although the overall structure of this tree is similar to that of S's tree, L's qualitative strategy is considerably more sophisticated. The qualitative constraints in the leaves of the tree are more complex and stronger. Obviously, L was paying attention to the load swing already at the initial stages of the task and making active effort at reducing the swing. Due to this, in actual (quantitative) execution traces, operator L was considerably more successful and was usually able to complete the task quicker than S. These speed-ups resulted from greater accelerations in the X direction.

6 Understanding control strategies in time

Qualitative trees above give good insight into tacit skill of operators. However, qualitative trees give static relations only, and a good intuition is still required from the user about the dynamics of the system to understand the mechanics of how a qualitative strategy achieves the goal. One remaining question is: Is it possible to generate, from a qualitative control strategy, explanation that involves behaviour in time? We look now at one approach to answering this.

The idea is to use a qualitative model of the dynamics of the controlled system and a qualitative control strategy, together with a qualitative simulator, to generate explanation of how the strategy works in time. We have explored this idea by using a Prolog implementation (Bratko 2001, chapter 20) of the qualitative simulation algorithm QSIM (Kuipers 1994). Qualitative simulation results provide a dynamic description of how an operator's control strategy achieves the goal of control.

To perform a qualitative simulation, we need a qualitative model of the modelled dynamic system. We used the following (approximate) qualitative model of the crane with fixed rope length, stated in terms of qualitative constraints used in QSIM:

deriv(X, DX)	Crane velocity DX is time derivative of crane position X
deriv(DX, DDX)	Crane acceleration DDX is time derivative of velocity DX
$DDX = M^+(F + M_0^+(Fi))$	Acceleration DDX is caused by control force F and the load
deriv(Fi, DFi)	Rope angular velocity DFi is derivative of angle Fi
deriv(DFi, DDFi)	Rope angular acceleration DDFi is derivative of DFi
$DDFi = M_0^-(F + M_0^+(Fi))$	Angular acceleration depends on force F and angle Fi

As an example, consider now the swing control with S's qualitative strategy near the goal position expressed as $DX = M_0^+(Fi)$. Let qualitatively the current state of the crane be such that the trolley's position X is positive and increasing, the trolley's velocity DX is positive and decreasing, the angle Fi is positive and decreasing, and angular velocity DFi is negative. Let the control task be to reach a steady state with all these four variables equal zero.

To enforce S's qualitative control constraint $DX = M^+(Fi)$, this qualitative constraint was simply added to the qualitative model of the crane as an additional constraint. The qualitative simulator was thus constrained to search within S's strategy, all the time correspondingly adjusting control force F. Notice that due to the branching in qualitative simulation (also including some spurious behaviours) several qualitative behaviours from the start to the goal state were thus generated. One of them is shown in Figure 5. It represents one way of explaining how at least one aspect of S's qualitative control strategy works. In the first part (the first two qualitative states), the control force is negative to change both the velocity DX and angle Fi from positive to zero and then to negative. When DX and Fi become negative, force F becomes positive to synchronously make both DX and Fi equal zero and steady.

7 Qualitative trees used as indirect controllers

Qualitative strategies in the form of qualitative constraints cannot be directly used for the (quantitative) computation of the next action. First, a quantitative constraint consistent with the qualitative constraint is needed, and this quantitative constraint can then be used in the concrete numerical calculation. There are many possible quantitative constraints consistent with the qualitative constraint and any of them corresponds to a candidate controller. So this approach defines an optimisation space over which an optimal controller can be sought. Since qualitative constraints are explicit, they also make it possible to take into account the knowledge of the control task when generating corresponding quantitative constraints. Experimental results in the crane domain (Šuc and Bratko 1999) are very convincing. The best human among the operator's traces (the best trace of the best among six humans) completes the control task in 51 seconds. The best clone qualitatively consistent with this operator's strategy does it in 38 seconds. Out of 90 controllers randomly generated without any background knowledge, 83 percent perform the task within the maximum allowed time set for human operators. When basic task knowledge was imposed to constrain the random generator of qualitatively consistent hypotheses, all 180 generated controllers performed the task within the time limit, with the mean time of 48 sec. (better than the best ever recorded human's time. This can be contrasted to experimental results with direct controllers. It has been experimentally found that in the crane domain, there is no more than 10 percent chance of inducing a successful direct controller (one that does the job within the time limit).

F	X	DX	Fi	DFi
neg/std	pos/inc	pos/dec	pos/dec	neg/std
neg/inc	pos/std	zero/dec	zero/dec	neg/inc
zero/std	pos/dec	neg/dec	neg/dec	neg/inc
pos/inc	pos/dec	neg/std	neg/std	zero/inc
pos/std	pos/dec	neg/inc	neg/inc	pos/inc
pos/dec	pos/dec	neg/inc	neg/inc	pos/std
pos/dec	pos/dec	neg/inc	neg/inc	pos/dec
zero/std	zero/std	zero/std	zero/std	zero/std

Figure 6: A qualitative behaviour satisfying S's control constraint $DX = M^+(Fi)$. Rows give successive actions (force F) and qualitative states of the dynamic system. Expressions of form Qval/Qdir are qualitative states of variables where Qval is a qualitative value and Qdir is a qualitative direction of change of a variable. For example, neg/std means that the value is negative (between minus infinity and zero), and it is steady; pos/inc means positive and increasing.

8 Conclusions

The problem of behavioural cloning of operators' skill has been discussed in the paper. When applying machine learning to this problem, two kinds of approaches were distinguished: direct and indirect controllers. Although more widely used, the induction of direct controllers clearly has difficulties. In contrast, the induction of indirect controllers enables a decomposition of the cloning task, which gives much more robust results and better controllers. This decomposition involves the concept of generalised operator's trajectory. Advantages are:

- The resulting controllers are much more robust than direct controllers.
- Qualitative abstraction of induced trajectories enables better explanation of the human's control strategy.
- It also allows the incorporation of explicit constraints that come from the definition of the control task.
- The qualitative abstraction defines a space for optimisation of induced controllers.

In this paper, indirect controllers in terms of qualitative constraints were discussed in particular. It was illustrated that such qualitative constraints induced from operators' control traces offer advantages particularly with respect to the understanding of human control strategies.

References

Bain, M., Sammut, C. (1999) In: *Machine Intelligence 15* (eds. K.Furukawa, D.Michie, S.Muggleton), Oxford University Press.

Bratko, I. (1997) Qualitative reconstruction of control skill. *Proc. QR97 (11th Int. Workshop on Qualitative Reasoning)*, pp. 41-52. Cortona, Italy, 1997 (Pubblicazioni N. 1036. Pavia: Instituto di analisi numerica.

Bratko, I. (2001) *Prolog Programming for Artificial Intelligence, 3rd edition.* Addison-Wesley.

I. Bratko, T. Urbančič (1999) Control skill, machine learning and hand-crafting in controller design. In: *Machine Intelligence 15* (eds. K.Furukawa, D.Michie, S.Muggleton), Oxford University Press.

I. Bratko, T. Urbančič, C. Sammut (1998) *Behavioural cloning of control skill In: Machine Learning and Data Mining: Methods and Applications* (eds. R.S.Michalski, I.Bratko, M.Kubat) Wiley.

Breiman, L., Friedman, J. H., Olshen, R. A., Stoe, C. J. (1984) *Classification and Regression Trees.* Belmont, CA: Wadswarth.

Karalič, A. (1992) Employing linear regression in regression tree leaves. *Proc. 10th European Conf. on Artificial Intelligence*, pp. 440-441. Vienna, 1992.

Karalič, A., Bratko, I. (1997) First Order Regression. *Machine Learning*, Vol. 26, 147-176.

Križman, V., Džeroski, S. and Kompare B. (1995) Discovering dynamics from measured data. *Electrotechnical Review*, Vol. 62, 19-198.

Kuipers, B. (1994) *Qualitative Reasoning: Modeling and Simulation with Incomplete Knowledge.* Cambridge, MA: MIT Press.

Michie, D. (1993) Knowledge, learning and machine intelligence. In: L.S.Sterling (ed.) *Intelligent Systems*, Plenum Press.

Michie, D., Bain, M., Hayes-Michie, J. (1990) Cognitive models from subcognitive skills. In: Grimble, M., McGhee, J., Mowforth, P. (eds.) *Knowledge-Based Systems in Industrial Control*, Stevenage: Peter Peregrinus.

Michie, D., Camacho, R. (1994) Building symbolic representations of intuitive real-time skills from performance data. In: K.Furukawa, S.Muggleton (eds.) *Machine Intelligence and Inductive Learning*, Oxford: Oxford University Press.

Quinlan, J. R. (1986) Induction of decision trees. *Machine Learning* **1**: 81-106..

Sammut, C. and Hurst, S. and Kedzier, D. and Michie, D. (1992) Learning to fly. Proc. *ICML '92 (9th Int. Conf. on Machine Learning)*, Aberdeen, 1992.

Šuc, D. (2001) *Machine Reconstruction of Human Control Strategies*. Ph. D. Thesis, Univ. of Ljubljana, Faculty of Computer and Information Sc.

Šuc, D., Bratko, I. (1997) Reconstructing control skill as LQ controllers with subgoals Proc. IJCAI'97, Yokohama, Japan, August 1997.

Šuc, D., Bratko, I. (1999) Symbolic and qualitative reconstruction of control skill. *ETAI Journal (Electronic Transactions on Artificial Intelligence)*, Vol. 3 (1999), Section B, pp. 1-22. http://www.ep.liu.se/ej/etai/1999/002/.

Šuc, D., Bratko, I. (2000a) Problem decomposition for behavioural cloning Proc. ECML'2000 (Europena Conf. Machine Learning), Barcelona, June 2000

Šuc, D., Bratko, I. (2000b)
Skill modeling through symbolic reconstruction of operator's trajectories. IEEE trans. syst. man cybern., Part A, Syst. humans, 2000, vol. 30, no. 6, str. 617-624.

Šuc, D., Bratko, I. (2000c) Qualitative trees applied to bicycle riding. *ETAI Journal (Electronic Transactions on Artificial Intelligence)*, Vol. 4 (2000), Section B, pp. 125-140. http://www.ep.liu.se/ej/etai/2000/014/.

Šuc, D., Bratko, I. (2001) Induction of qualitative trees. ECML'2000 (Europena Conf. Machine Learning), Barcelona, June 2000

Urbančič, T., Bratko, I. (1994) Reconstructing human subcognitive skill through machine learning. *Proc. ECAI-94*, pp. 498-502. Amsterdam, 1994.

Local Structure Learning in Graphical Models

Christian Borgelt and Rudolf Kruse
Dept. of Knowledge Processing and Language Engineering
Otto-von-Guericke-University of Magdeburg
Universitätsplatz 2, D-39106 Magdeburg, Germany
e-mail: {borgelt,kruse}@iws.cs.uni-magdeburg.de

Abstract A topic in probabilistic network learning is to exploit local network structure, i.e., to capture regularities in the conditional probability distributions, and to learn networks with such local structure from data. In this paper we present a modification of the learning algorithm for Bayesian networks with a local decision graph representation suggested in Chickering et al. (1997), which is often more efficient. It rests on the idea to exploit the decision graph structure not only to capture a larger set of regularities than decision trees can, but also to improve the learning process. In addition, we study the influence of the properties of the evaluation measure used on the learning time and identify three classes of evaluation measures.

1 Introduction

Probabilistic inference networks—especially Bayesian networks Pearl (1992), but also Markov networks Lauritzen and Spiegelhalter (1988)—are well-known tools for reasoning under uncertainty in multidimensional spaces. The idea underlying them is to exploit independence relations between variables in order to decompose a multivariate probability distribution into a set of (conditional or marginal) distributions on lower-dimensional subspaces. Efficient implementations include HUGIN Andersen et al. (1989) and PATHFINDER Heckerman (1991).

Such independence relations have been studied extensively in the field of graphical modeling Kruse et al. (1991) and though using them to facilitate reasoning in multidimensional domains has originated in probabilistic reasoning, this approach has been generalized to be usable with other uncertainty calculi Shafer and Shenoy (1988), e.g. in the so-called valuation-based networks Shenoy (1991), and has been implemented e.g. in PULCINELLA Saffiotti and Umkehrer (1991).

Due to their connection to fuzzy systems and their ability to deal not only with uncertainty but also with imprecision, recently possibilistic networks also gained some attention Gebhardt (1997); Borgelt and Kruse (2002). They have been implemented e.g. in POSSINFER Gebhardt and Kruse (1995a); Kruse et al. (1994). In this paper we consider Bayesian networks and a type of possibilistic networks that is based on the context-model interpretation of a degree of possibility Gebhardt and Kruse (1993).

A Bayesian network is a directed acyclic graph in which each node represents a variable that is used to describe some domain of interest, and each edge represents a direct dependence between two variables. The structure of the directed graph encodes a set of conditional independence statements that can be read from the graph using a graph theoretic criterion called *d-separation* Pearl (1992). In addition, it represents a particular joint probability distribution, which is specified by assigning to each node in the network a (conditional) probability distribution for the values of the corresponding variable given the parent variables in the network (if any).

Formally, a Bayesian network describes a factorization of a multivariate probability distribution that results from an application of the product theorem of probability theory to the joint distribution and a simplification of the factors achieved by exploiting conditional independence statements of the form $P(A \mid B, X) = P(A \mid X)$, where A and B are variables and X is a set of variables. Hence, the represented joint distribution can be computed as

$$P(A_1, \ldots, A_n) = \prod_{i=1}^{n} P(A_i \mid \text{parents}(A_i)),$$

where $\text{parents}(A_i)$ is the set of parents of variable A_i.

The directed acyclic graph of a Bayesian network captures the *global structure* of the underlying domain, i.e., the structure of (conditional) dependences and independences, but fails to take into account *local structure* that may be present in the conditional probability distributions stored with the nodes. An important issue in Bayesian network research is to capture such local structure and enable learning it from data.

In this paper we present a modification of the approach presented in Chickering *et al.* (1997) to learn Bayesian networks with a local decision graph structure from data. Our approach rests on exploiting the decision graph structure not only to capture a larger set of regularities in conditional probability tables but also to simplify the learning process. Our approach is also more efficient, because it needs fewer visits to the database to learn from.

Furthermore, we apply our local structure learning method to learning possibilistic networks from data. The transfer to this type of networks is straightforward. We also consider a large variety of evaluation measures (or scoring functions) for both probabilistic and possibilistic network learning. Many of these measures originated from decision tree learning, but can also be applied to learning Bayesian networks if the parents of a variable in a Bayesian network are seen as combined into one pseudo-variable. Some of them can easily be transferred to the possibilistic case. We study the influence of the evaluation measure on the running time of the learning algorithm and identify three classes of evaluation measures. Finally, we present experimental results for both learning Bayesian networks and possibilistic networks.

2 Possibilistic Networks

The development of possibilistic networks was triggered by the fact that probabilistic networks are well suited to represent and process *uncertain* information, but cannot that easily be extended to handle *imprecise* information. Since the explicit treatment of imprecise information is more and more claimed to be necessary for industrial practice, it is reasonable to investigate graphical models related to alternative uncertainty calculi, e.g. possibility theory.

Maybe the best way to explain the difference between uncertain and imprecise information is to consider the notion of a degree of possibility. The interpretation we prefer is based on the context model Gebhardt and Kruse (1993); Kruse *et al.* (1994). In this model possibility distributions are seen as information-compressed representations of (not necessarily nested) random sets and a degree of possibility as the one-point coverage of a random set Nguyen (1984).

Let ω_0 be the actual, but unknown state of a domain of interest, which is contained in a set Ω of possible states. Let $(C, 2^C, P)$, $C = \{c_1, c_2, \ldots, c_m\}$, be a finite probability space and $\gamma : C \to 2^\Omega$ a set-valued mapping. C is seen as a set of contexts that have to be distinguished for a set-valued specification of ω_0. The contexts are supposed to describe different physical and observation-related frame conditions. $P(\{c\})$ is the (subjective) probability of the (occurrence or selection of the) context c.

A set $\gamma(c)$ is assumed to be the *most specific correct set-valued specification* of ω_0, which is implied by the frame conditions that characterize the context c. By 'most specific set-valued specification' we mean that $\omega_0 \in \gamma(c)$ is guaranteed to be true for $\gamma(c)$, but is not guaranteed for any proper subset of $\gamma(c)$. The resulting *random set* $\Gamma = (\gamma, P)$ is an imperfect (i.e. imprecise *and* uncertain) specification of ω_0. Let π_Γ denote the *one-point coverage of* Γ (the *possibility distribution induced by* Γ), which is defined as

$$\pi_\Gamma : \Omega \to [0,1], \pi_\Gamma(\omega) = P(\{c \in C \mid \omega \in \gamma(c)\}).$$

In a complete modeling, the contexts in C must be specified in detail, so that the relationships between all contexts c_j and their corresponding specifications $\gamma(c_j)$ are made explicit. But if the contexts are unknown or ignored, then $\pi_\Gamma(\omega)$ is the total mass of all contexts c that provide a specification $\gamma(c)$ in which ω_0 is contained, and this quantifies the *possibility of truth* of the statement "$\omega = \omega_0$" Gebhardt and Kruse (1993, 1996).

That in this interpretation a possibility distribution represents uncertain *and* imprecise knowledge can be understood best by comparing it to a probability distribution and to a relation. A probability distribution covers *uncertain*, but *precise* knowledge. This becomes obvious, if one notices that a possibility distribution in the interpretation described above reduces to a probability distribution, if $\forall c_j \in C : |\gamma(c_j)| = 1$, i.e. if for all contexts the specification of ω_0 is precise. On the other hand, a relation represents *imprecise*, but *certain* knowledge about dependences between attributes. Thus, not surprisingly, a relation can also be seen as a special case of a possibility distribution, namely if there is only one context. Hence the context-dependent specifications are responsible for the imprecision, the contexts for the uncertainty in the imperfect knowledge expressed by a possibility distribution.

With this interpretation the theory of possibilistic networks can be developed in analogy to the probabilistic case. The only difference is that instead of the product to determine a new joint distribution and the sum to determine a (new) marginal distribution, the operations minimum and maximum have to be used.

As a concept of possibilistic independence we use possibilistic non-interactivity. Let X, Y, and Z be three disjoint subsets of variables. Then X is called *conditionally independent* of Y given Z w.r.t. π, if $\forall \omega \in \Omega$:

$$\pi(\omega_{X \cup Y} \mid \omega_Z) = \min\{\pi(\omega_X \mid \omega_Z), \pi(\omega_Y \mid \omega_Z)\}$$

whenever $\pi(\omega_Z) > 0$, where $\pi(\cdot \mid \cdot)$ is a non-normalized conditional possibility distribution

$$\pi(\omega_X \mid \omega_Z) = \max\{\pi(\omega') \mid \omega' \in \Omega \wedge \operatorname{proj}_X(\omega) = \omega_X \wedge \operatorname{proj}_Z(\omega) = \omega_Z\},$$

with $\operatorname{proj}_X(\omega)$ the projection of a tuple ω to the variables in X.

Learning possibilistic networks from data has been studied in Gebhardt and Kruse (1995b, 1996); Borgelt and Kruse (1997a,b). The idea to exploit local structure can be applied directly to (conditional) possibility distributions, since it is not bound to any specific uncertainty or imprecision calculus.

3 Local Network Structure

Whereas the global structure of a probabilistic or possibilistic network is the directed acyclic graph that encodes the conditional independence statements that hold in a certain domain of interest, the term "local structure" refers to regularities in the conditional probability or possibility tables that are stored with the nodes of the network. Several approaches to exploit such regularities have been studied for Bayesian networks in order to capture additional (i.e. context specific) independences and thus to (potentially) enhance inference. Among these are similarity networks Heckerman (1991) and the related multinets Geiger and Heckerman (1991), the use of asymmetric representations for decision making Smith *et al.* (1993) and probabilistic Horn rules Poole (1993), and finally also decision trees Boutilier *et al.* (1996) and decision graphs Chickering *et al.* (1997). In this paper we focus on the decision tree/decision graph approach, since it appears to be the most convenient one, and review it in the following for discrete Bayesian networks (i.e., in which all variables are discrete).

A very simple way to encode a conditional probability distribution is a table, which for each combination of values of the conditioning variables contains a line stating the corresponding conditional probability distribution for the values of the conditioned variable. As a simple example, let us consider the small section of a Bayesian network shown in figure 1 (and let us assume that in this network the variable C has no other parents than variables A and B). Let $\operatorname{dom}(A) = \{a_1, a_2, a_3\}$, $\operatorname{dom}(B) = \{b_1, b_2\}$, and $\operatorname{dom}(C) = \{c_1, c_2\}$. Then the conditional probabilities $P(C = c_k \mid A = a_i, B = b_j)$ have to be stored with the node for variable C, e.g. as shown in table 1. The second column contains only entries $1 - p_i$, because the probabilities have to sum to 1 and there are only two possible values for variable C.

The same conditional probability distribution can also be stored in a tree, where the leaves hold the conditional probability distributions and each level of inner nodes corresponds to one conditioning variable (see figure 2). The branches in this tree are labeled with the values of the conditioning variables and thus each path from the root to a leaf corresponds to one combination of values of the conditioning variables. Obviously such a tree is equivalent to a decision tree for the variable C (like one learned e.g. by

Figure 1. A small section of a Bayesian network.

parents		child	
A	B	$C = c_1$	$C = c_2$
a_1	b_1	p_1	$1 - p_1$
a_1	b_2	p_2	$1 - p_2$
a_2	b_1	p_3	$1 - p_3$
a_2	b_2	p_4	$1 - p_4$
a_3	b_1	p_5	$1 - p_5$
a_3	b_2	p_6	$1 - p_6$

Table 1. A conditional probability table for the network section shown in figure 1.

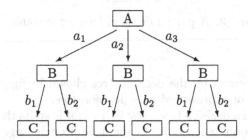

Figure 2. A full decision tree for variable C.

the well-known decision tree induction program C4.5 Quinlan (1993)) with the following restrictions: All leaves have to lie on the same level and in one level of the tree the same variable has to be tested on all paths. If these restrictions hold, we call the tree a *full* decision tree, because all possible combinations of values of the test attributes are explicitly represented in the tree.

Let us assume now that there are some regularities in the conditional probability distribution (see table 2), that is, let certain conditional probabilities be identical. Since the table clearly shows that the value of the variable B is important only if A has the value a_2, the tests of variable B can be removed from the branches for the values a_1 and a_3 (see figure 3). This shows the advantages of a decision tree representation.

Unfortunately, however, a decision tree is not powerful enough to capture all possible regularities that may be present in a conditional probability table. Although we can achieve a lot by accepting a change in the test order of the variables and by accepting binary splits and multiple tests of the same variable (then, for example, the regularities

parents		child	
A	B	$C = c_1$	$C = c_2$
a_1	b_1	p_1	$1 - p_1$
a_1	b_2	p_1	$1 - p_1$
a_2	b_1	p_3	$1 - p_3$
a_2	b_2	p_4	$1 - p_4$
a_3	b_1	p_2	$1 - p_2$
a_3	b_2	p_2	$1 - p_2$

Table 2. A conditional probability table for the section of a Bayesian network shown in figure 1 with some regularities.

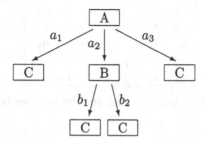

Figure 3. A partial decision tree for variable C.

in table 3 can be represented by the decision tree shown in figure 4), the regularities shown in table 4 cannot be represented by a decision tree.

The problem is that in a decision tree a test of a variable splits the lines of a conditional probability table into disjoint subsets that cannot be brought together again. In table 4 a test of variable B thus separates lines 1 and 2 and a test of variable A separates lines 4 and 5. Hence either test prevents us from exploiting one of the two equivalences of probabilities. This drawback can be overcome by allowing a node of the tree to have more than one parent, thus going from decision trees to decision graphs Chickering *et al.* (1997). With decision graphs the regularities in table 4 can easily be captured, see figure 5.

4 Learning Local Structure

To learn a decision graph three operations are defined in Chickering *et al.* (1997):
- *full split*: Split a leaf node according to the values of some variable.
- *binary split*: Split a leaf node such that one child corresponds to some value a_k of some variable and the other child to all other values of this variable.
- *merge*: merge two distinct leaf nodes.

A greedy algorithm based on these operations can easily be found Chickering *et al.* (1997). It applies all possible operations of the types defined above to a given decision

parents		child	
A	B	$C = c_1$	$C = c_2$
a_1	b_1	p_1	$1 - p_1$
a_1	b_2	p_1	$1 - p_1$
a_2	b_1	p_2	$1 - p_2$
a_2	b_2	p_3	$1 - p_3$
a_3	b_1	p_2	$1 - p_2$
a_3	b_2	p_4	$1 - p_4$

Table 3. A conditional probability table for the section of a Bayesian network shown in figure 1 with a second kind of regularities.

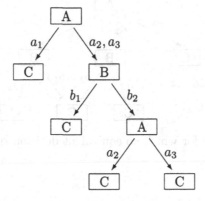

Figure 4. A decision tree with two tests of variable A that captures the regularities in the conditional probability table shown in table 3.

graph and then selects that operation (if any) that leads to the highest improvement of the network score. This search is carried out until no operation can be found that leads to an improvement.

Our approach is only a slight modification of the above. The additional degree of freedom of decision graphs compared to decision trees, namely that a node in a decision graph can have more than one parent, can be exploited not only to capture a larger set of regularities but also to improve the learning process for the local structure of a Bayesian network. Our idea is as follows: With decision graphs, we can always work with the complete set of inner nodes of a full decision tree and let only leaves have more than one parent. Even if we do not care about the order of the conditioning variables in the decision structure and allow only one test per variable on each path, such a structure can capture all regularities in the examples examined in the preceding section. For example, the regularities of table 3 are captured by the decision graph with a full set of inner nodes shown in figure 6.

parents		child	
A	B	$C = c_1$	$C = c_2$
a_1	b_1	p_1	$1 - p_1$
a_1	b_2	p_1	$1 - p_1$
a_2	b_1	p_2	$1 - p_2$
a_2	b_2	p_3	$1 - p_3$
a_3	b_1	p_3	$1 - p_3$
a_3	b_2	p_4	$1 - p_4$

Table 4. A conditional probability table for the section of a Bayesian network shown in figure 1 with a third kind of regularities.

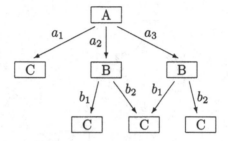

Figure 5. A decision graph for which no equivalent decision tree exists. It captures the regularities in table 4

It is easy to see that such an approach can capture any regularities that may be present in conditional probability tables. Basically, merging the leaves of a full decision tree is the same as merging lines of a conditional probability table. The decision graph structure just makes it much easier to keep track of the different value combinations of the conditioning (i.e. parent) variables, for which the same probability distribution for the values of the conditioned (i.e. child) variable has to be adopted.

In a learning algorithm we use only two operations, namely (1) adding a new level to a decision tree/graph, i.e., splitting all leaves according to the values of a new parent variable, and (2) merging two leaves into one. The first step, which may seem to be costly, does no harm, since it is necessary, even if one only learns a Bayesian network without local structure (provided the conditional distributions are represented as a decision tree). Only this step involves visiting the database to learn from in order to determine the conditional value frequencies. The next step, in which leaves are merged, can be carried out without visiting the database, since all necessary information is already available in the leaf nodes (provided the original leaf nodes are kept during a trial merge and are simply restored afterwards). Thus we need to visit the database only as often as an algorithm for learning a Bayesian network without local structure does. In contrast to this, the algorithm presented in Chickering et al. (1997) needs to visit the database each time a split of leaf nodes is considered. This can exceed by far the number of times

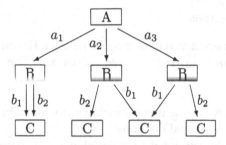

Figure 6. A decision graph with a full set of inner nodes that captures the regularities in table 3. Note that the test of variable B in the leftmost node on the second level is without effect, because both edges lead to the same leaf.

an algorithm for learning a network without local structure needs to visit the database, especially, since multiple tests of the same variable along the same path are permitted.

The leaf merging process is often less costly as it may seem at first sight, since we can exploit the fact that several evaluation measures (or scoring functions) are computed leaf by leaf or from terms that are computed leaf by leaf (see below). Hence, when two leaves are merged, the decision graph need not be reevaluated completely, but the change can often be computed locally from the frequency distributions in the merged leaves and the distribution in the resulting leaf.

To find the best set of mergers of leaf nodes, one can use any of the well-known search heuristics, e.g. greedy search or, if a mechanism for re-splitting leaf nodes is provided (which is easy to program), simulated annealing. Since we chose a greedy parent selection on a topological order (that is, the well-known K2 search method Cooper and Herskovits (1992)) in our experiments, we implemented a simple greedy search. That is, we always merge those two leaf nodes, that lead to the highest improvement of the evaluation measure. The merging process stops, if no leaf merger improves the value of the evaluation measure. However, we implemented the greedy merging in two different ways. In the first approach any merger between two leaf nodes of the current decision graph can be selected. We call this *unrestricted merging*. In the second approach, we first consider merging only such leaves that have the same parent. Only after no merger improving the evaluation measure can be found anymore, we allow mergers of leaves that have the same grandparent, and so on. This approach we call *levelwise merging*, since we climb up in the tree level by level to determine which leaves are considered for a possible merger. The latter approach can be slightly more efficient, since in general a slightly smaller set of mergers is considered. It can also lead to a simpler structure, since mergers of leaves that are "far apart" in the tree are less likely.

Of course, our approach can result in a complicated structure that may hide a simple structure of context-specific independences. But the same is true, though maybe less likely, for the algorithm presented in Chickering *et al.* (1997) and thus some postprocessing to simplify the structure found by the algorithm—for instance, by changing the order of variables and by splitting tests along a path—is always advisable.

5 Evaluation Measures

The process of selecting parent variables when learning a Bayesian network is very similar to selecting a test attribute in decision tree induction. The only difference is that in decision tree learning only single attributes are considered, whereas in Bayesian network learning there can be more than one parent. But this is not really a difference, since we can always view all parents as one pseudo-variable, the domain of which is the Cartesian product of the parents' individual domains.

This view can easily be extended to a decision graph representation, where several paths (and thus several pseudo-values) can lead to the same leaf (the same conditional probability distribution). In this case we only have to combine certain elements of the Cartesian product of the parents' domains into one pseudo-element. For example, for the decision graph shown on figure 5, we can view the parents A and B as one pseudo-variable X with $\text{dom}(X) = \{x_1, x_2, x_3, x_4\}$, where $x_1 \hat{=} (a_1, b_1) \vee (a_1, b_2)$, $x_2 \hat{=} (a_2, b_1)$, $x_3 \hat{=} (a_2, b_2) \vee (a_3, b_1)$ and $x_4 \hat{=} (a_3, b_2)$.

The only thing we have to take care of is that in contrast to the measures commonly used for Bayesian network learning, like Bayesian measures based on the Bayesian Dirichlet metric or measures based on the minimum description length principle, attribute selection measures for decision tree induction usually do not have a built-in property that prevents them from selecting too many parent variables. An example is information gain, which for decision tree induction is known to be biased towards many-valued attributes.[1] Since an additional parent variable obviously increases the number of values of the pseudo-attributes, information gain tends to select too many parents. Fortunately, this drawback can easily be overcome by requiring a candidate parent to improve the value of an evaluation measure by a predefined minimal amount, before this candidate is considered eligible. We made this parameter an optional argument of our program.

Limits of space prevent us from describing in detail the evaluation measures we used in the experiments reported in section 6. Hence we only list them here without much explanation. An interested reader is asked to consult the references or Borgelt and Kruse (2002), which discusses them in some detail.

Probabilistic Measures

- information gain I_{gain} Kullback and Leibler (1951); Chow and Liu (1968) (mutual information/cross entropy)
- information gain ratio I_{gr} Quinlan (1993)
- symmetric information gain ratio I_{sgr} Lopez de Mantaras (1991)
- Gini index Breiman *et al.* (1984); Wehenkel (1996)
- symmetric Gini index Zhou and Dillon (1991)
- modified Gini index Kononenko (1994)
- relief measure Kira and Rendell (1992); Kononenko (1994)
- relevance Baim (1988)

---

[1] The reason is that a split of a value of a test attribute into two values can lead only to the same or a higher information gain, and in practice almost always leads to a higher information gain, mainly due to a quantization effect.

- χ^2 measure
- K2 metric Cooper and Herskovits (1992); Heckerman *et al.* (1995)
- BDeu metric Buntine (1991); Heckerman *et al.* (1995)
- minimum description length with coding based on relative frequencies l_{rel} Kononenko (1995)
- minimum description length with coding based on absolute frequencies l_{abs} Kononenko (1995) (closely related to the K2-metric)
- stochastic complexity Krichevsky and Trofimov (1983); Rissanen (1987)

Probabilistic Measures

- possibilistic analog of the χ^2-measure Borgelt and Kruse (1997a)
- possibilistic analog of mutual information (mutual specificity) Borgelt and Kruse (1997a)
- specificity gain S_{gain} Gebhardt and Kruse (1996); Borgelt and Kruse (1997a)
- specificity gain ratio S_{gr} Borgelt and Kruse (1997a)
- symmetric specificity gain ratio S_{sgr} Borgelt and Kruse (1997a)

When it comes to learning the local structure of a graphical model, it becomes important whether an evaluation measure can be computed from individual terms for each of the leaves of the decision graph representing the conditional distribution to assess, which makes it possible to compute the new value of the measure after merging two leaves by computing a simple delta, or whether it is not possible to find the new value by such "local" computations, so that the whole conditional distribution has to be reevaluated. This consideration leads to three classes of evaluation measures:

- The improvement resulting from a merger is independent of other mergers.

  Examples: o χ^2 measure
             o   information gain
             o   K2 metric

- The improvement resulting from a merger depends on other mergers, but can be computed locally from the merged leaves and certain cached values.

  Examples: o   information gain ratio
             o   symmetric/modified Gini index

- The improvement resulting from a merger depends on other mergers in such a way that the full tree has to be reevaluated.

  Examples: o   specificity gain
             o   (symmetric) specificity gain ratio

In order to understand this distinction, let us briefly take a closer look at one example from each class. For the first class we consider the *K2 metric* Cooper and Herskovits (1992), which is based on a Bayesian approach. The idea underlying it is to compute the probability of a (directed) graph structure given the data, i.e., to compute

$$P(\vec{G} \mid D) = \frac{1}{P(D)} \int_{\Theta} P(D \mid \vec{G}, \Theta) f(\Theta \mid \vec{G}) P(\vec{G}) \, \mathrm{d}\Theta,$$

where $\vec{G}$ is the directed graph underlying the graphical model, D is the dataset to learn from, and Θ is the set of parameters of the model, i.e., the conditional probabilities. f describes the prior probability (in a Bayesian sense) of a each assignment of parameter values given the structure of the graph. By restricting our considerations to a Bayes factor for comparing networks, which eliminates the need to explicitly compute the probability of the database, and by making certain assumptions about data and parameter independence Cooper and Herskovits (1992), we get

$$P(\vec{G}, D) = \gamma \prod_{k=1}^{r} \prod_{j=1}^{m_k} \int_{\theta_{ijk}} \cdots \int \left(\prod_{i=1}^{n_k} \theta_{ijk}^{N_{ijk}} \right) f(\theta_{1jk}, \ldots, \theta_{n_k jk}) \, d\theta_{1jk} \ldots d\theta_{n_k jk},$$

where γ is a normalization factor, r is the number of variables, m_k the number of distinct instantiations of the parent variables of variable k, n_k the number of values of variable k, θ_{ijk} the conditional probability that variable k has the i-th value given that its parent variables are instantiated with the j-th combination of values, and N_{ijk} is the number of times variable k is instantiated with its i-th value and its parents are instantiated with their j-th value combination in the database D to learn from. To solve this formula, $f(\theta_{1jk}, \ldots, \theta_{n_k jk}) = $ const. is chosen Cooper and Herskovits (1992) and then the solution can be obtained with Dirichlet's integral:

$$K_2(\vec{G}, D) = \gamma \prod_{k=1}^{r} \prod_{j=1}^{m_k} \frac{(n_k - 1)!}{(N_{.jk} + n_k - 1)!} \prod_{i=1}^{n_k} N_{ijk}!.$$

In implementations the logarithm of this measure is computed, so that the products turn into sums. From this formula it is obvious, that merging two leaves removes two factors from the second product, namely the two that refer to the two merged leaves, and adds a new one that refers to the result of the merger. Therefore the change of this measure as it results from merging leaves can easily be computed as a simple delta. This makes the computations very efficient.

As an example of an evaluation measure from the second class we consider the *information gain ratio* Quinlan (1993). This measure is based on Shannon entropy $H = -\sum_{i=1}^{n} p_i \log_2 p_i$ and can be seen as a normalization of the *information gain*

$$
\begin{aligned}
I_{\text{gain}}(A, B) \quad &= \quad H_A - H_{A|B} \;=\; H_A + H_B - H_{AB} \\[1em]
&= \quad - \sum_{a \in \text{dom}(A)} P(A = a) \log_2 P(A = a) \\[1em]
&\quad - \sum_{b \in \text{dom}(B)} P(B = b) \log_2 P(B = b) \\[1em]
&\quad + \sum_{a \in \text{dom}(A)} \sum_{b \in \text{dom}(B)} P(A = a, B = b) \log_2 P(A = a, B = b),
\end{aligned}
$$

namely as

$$I_{\text{gr}}(A, B) \quad = \quad \frac{I_{\text{gain}}(A, B)}{H_A} \quad = \quad \frac{H_A + H_B - H_{AB}}{H_A}$$

The normalization is meant to remove the abovementioned bias towards many-valued attributes. It is easy to see that information gain allows to compute the change that results from merging two leaves as a simple delta, since only terms from the first and the third sum have to be replaced, while it is not possible to compute such a delta for information gain ratio due to the normalization factor. However, if we cache the values of the entropies it is computed from, the recomputation involves almost no additional costs. The entropies can be adapted by computing a delta resulting from the merger and then we only have to recompute the quotient.

As an example of an evaluation measure from the third class we consider the *specificity gain* Gebhardt and Kruse (1996); Borgelt and Kruse (1997a). It can be seen as a generalization of Hartley information gain on the basis of an α-cut view of possibility distributions and is defined as

$$
S_{\text{gain}}(A, B) = \int_0^{\sup \Pi} \log_2 \left(\sum_{a \in \text{dom}(A)} [\Pi]_\alpha(A = a) \right) + \log_2 \left(\sum_{b \in \text{dom}(B)} [\Pi]_\alpha(B = b) \right)
$$
$$
- \log_2 \left(\sum_{a \in \text{dom}(A)} \sum_{b \in \text{dom}(B)} [\Pi]_\alpha(A = a, B = b) \right) d\alpha.
$$

The formula shows that this measure is analogous to information gain. However, it does not share all of the nice properties of information gain. In particular, its change as it results from merging two leaves cannot be computed as a simple delta. The reason is that the computation of this measures involves sorting the degrees of possibility, and if two leaves are merged, they have to be resorted. This also makes it clear why no values can be cached to make a local computation possible. Fortunately, only the specificity based measures have this disadvantageous property. All other measures, possibilistic as well as probabilistic, belong to one of the other two classes.

6 Experimental Results

All experiments reported here were carried out with a prototype learning program for probabilistic and possibilistic networks called INES (Induction of NEtwork Structures, written by the first author of this paper), into which the described method and all of the listed evaluation measures are incorporated. This program as well as datasets and shell scripts to carry out the experiments can be retrieved free of charge at

http://fuzzy.cs.uni-magdeburg.de/~borgelt/software.html#ines

As a test case we chose the Danish Jersey cattle blood group determination problem Rasmussen (1992), for which a Bayesian network designed by domain experts and a database of 500 real world sample cases exist. Nevertheless, for learning Bayesian networks, we did not use the real world database, since it contains a lot of missing values. Instead, we used 20 artificially generated databases with 1000 sample cases each, 10 of which we used for learning, 10 for testing the learning result, over which the results were then averaged. The real world dataset was used only for learning possibilistic networks, since with them, missing values can be handled directly.

eval. measure	num. of conds.	add. conds.	miss. conds.	num. of params.	network quality train	test
indep. vars.	0.0	0.0	22.0	59	-19921	-20087
original	22.0	0.0	0.0	219	-11391	-11506

Table 5. Reference evaluations for Bayesian network learning.

eval. measure	num. of conds.	add. conds.	miss. conds.	num. of params.	network quality train	test
I_{gain}	35.0	17.1	4.1	1342	-11229	-11818
I_{gr}	24.0	6.7	4.7	209	-11615	-11737
I_{sgr}	32.0	11.3	1.3	317	-11388	-11575
Gini	35.0	17.1	4.1	1342	-11233	-11813
χ^2	35.0	17.3	4.3	1301	-11235	-11805
K2 metric	23.3	1.4	0.1	230	-11385	-11512
BDeu metric	31.2	9.3	0.1	276	-11385	-11521
l_{rel}	22.5	0.6	0.1	220	-11390	-11508

Table 6. Results of Bayesian network learning without local structure.

To evaluate the quality of the learned network, we chose the following approach: Given a Bayesian network, the probability of any (complete) sample case can easily be computed. If we assume the sample cases to be independent, we can compute from these probabilities the probability of the whole database (simply as their product). If we assume all network structures to have the same prior probability, this database probability is a direct measure of the network quality.

For possibilistic networks, we used a similar approach. Given a possibilistic network, the possibility degree of any (complete) tuple can be computed. If a tuple contains missing values, we assign to this tuple the maximal possibility degree over all complete tuples that are compatible with this tuple. The sum of these possibility degrees we used as a quality measure. This is justified, since due to the the way in which a possibilistic network approximates a multivariate possibility distribution, the possibility degree resulting from the network must always be equal or greater than the true possibility degree. Hence, the lower the sum of the possibility degrees for the tuples in the database, the better the network. More details about this evaluation method can be found in Borgelt and Kruse (1997b, 2002).

The results of some of our experiments are shown in tables 5 to 13. In addition to the network evaluation, these tables show the total number of conditions (parents) as a measure of the complexity of the global network structure, the number of additional and missing edges compared to the human expert designed reference network (which is reasonable only for Bayesian network learning), and the number of (probability or possibility) parameters as a measure of the complexity of the local network structure.

eval. measure	num. of conds.	add. conds.	miss. conds.	num. of params.	network quality train	test
I_{gain}	35.0	17.1	4.1	1260	−11192	−11806
I_{gr}	31.6	11.0	1.4	133	−14979	−15151
I_{sgr}	34.7	13.9	1.2	342	−11424	−11675
Gini	35.0	17.1	4.1	1254	−11195	−11802
χ^2	35.0	17.3	4.3	1216	−11197	−11794
K2 metric	26.4	4.5	0.1	195	−11341	−11507
BDeu metric	36.0	14.3	0.3	306	−11336	−11505
l_{rel}	25.1	3.8	0.7	219	−11350	−11498

Table 7. Results of Bayesian network learning with local structure (unrestricted).

eval. measure	num. of conds.	add. conds.	miss. conds.	num. of params.	network quality train	test
I_{gain}	35.0	17.1	4.1	1260	−11192	−11806
I_{gr}	32.1	11.7	1.6	132	−15202	−15354
I_{sgr}	34.7	13.9	1.2	342	−11424	−11675
Gini	35.0	17.1	4.1	1254	−11195	−11802
χ^2	35.0	17.3	4.3	1216	−11197	−11794
K2 metric	26.3	4.4	0.1	195	−11341	−11508
BDeu metric	35.9	14.2	0.3	305	−11338	−11504
l_{rel}	25.0	3.7	0.7	219	−11350	−11497

Table 8. Results of Bayesian network learning with local structure (levelwise).

eval. measure	num. of conds.	add. conds.	miss. conds.	num. of params.	network quality train	test
I_{gain}	35.0	17.1	4.1	1260	−11192	−11806
I_{gr}	24.0	6.7	4.7	121	−14752	−14926
I_{sgr}	32.0	11.3	1.3	217	−11452	−11650
Gini	35.0	17.1	4.1	1253	−11195	−11802
χ^2	35.0	17.3	4.3	1216	−11197	−11794
K2 metric	23.3	1.4	0.1	168	−11352	−11492
BDeu metric	31.2	9.3	0.1	218	−11357	−11484
l_{rel}	22.5	0.6	0.1	162	−11357	−11488

Table 9. Results of Bayesian network learning with local structure preserving the global structure.

Table 5 shows the evaluation results for a graph without edges (independent variables) and the human expert designed reference structure. These evaluation can be used as a baseline for comparisons. From table 6 it can be seen that some measures tend to select too many conditions (parents), thus leading to overfitting. As already said, this disadvantage can be amended to some degree by requiring a certain minimal improvement of the network evaluation when adding a condition.

At first sight it is surprising that allowing local structure to be learned (see tables 7 to 9), although in most cases it leads to a reduction of the number of necessary parameters, makes the global structure more complex, since for several measures the number of selected conditions is larger than for networks without local structure. But a second thought (and a closer inspection of the learned networks) reveals that this could have been foreseen. In a frequency distribution determined from a database of sample cases random fluctuations are to be expected. Usually these do not lead to additional conditions (except for measures like information gain), since the "costs" of an additional level with several (approximately) equivalent leaves prevents the selection of such a condition. But the disadvantage of (approximately) equivalent leaves is removed by the possibility to merge these leaves, and thus those fluctuations that show a higher deviation from the true (independent) probability distribution are filtered out and become significant to the measure. Information gain ratio seems to be an especially pronounced example. The effect occurs for both unrestricted and levelwise merging, which lead to very similar results.

This effect reduces when learning from a larger dataset, but does not vanish completely. We believe this to be a general problem any learning algorithm for local structure has to cope with. Therefore it may be advisable not to combine learning global and local network structure, but to learn the global structure first and to simplify the learned structure afterwards by learning the local structure. To check this assumption, we applied learning the local structure to the outcome of global structure learning, with the sets of parents fixed. The result, which is shown in table 9, is indeed slightly better. However, information gain ratio still yields very bad results compared to the other measures, thus indicating that it is not adequate to select the leaves to merge and to determine when to stop merging.

The results of learning possibilistic networks with local structure, which are shown in tables 10 to 14, are very similar to the results of probabilistic network learning. However, the gains from local structure learning while preserving the learned global structure seem to be much smaller here and thus it seems to be more advisable to combine local and global structure learning.

7 Conclusions

In this paper we presented a method to learn the local structure of a Bayesian network from data, which we believe to be more efficient than the approach presented in Chickering et al. (1997). We applied the same idea to possibilistic networks, thus arriving at an algorithm to learn possibilistic networks with local structure. The experimental results show that trying to learn local structure has to be handled with care, since it can lead to the counter-intuitive effect of a more complicated global structure. Maybe it is advisable

eval. measure	num. of conds.	num. of params.	network quality		
			avg.	min.	max.
indep. vars.	0	80	10.160	10.064	11.390
original	??	308	9.917	9.888	11.318

Table 10. Results of possibilistic network learning without local structure.

eval. measure	num. of conds.	num. of params.	network quality		
			avg.	min.	max.
S_{gain}	31	1630	8.621	8.524	10.292
S_{gr}	18	196	9.553	9.390	11.100
S_{sgr}	28	496	9.057	8.946	10.740
poss. χ^2	35	1486	8.329	8.154	10.200
mut. spec.	33	774	8.344	8.206	10.416

Table 11. Results of possibilistic network learning without local structure.

eval. measure	num. of conds.	num. of params.	network quality		
			avg.	min.	max.
S_{gain}	34	768	8.739	8.548	10.620
S_{gr}	21	215	9.637	9.450	11.254
S_{sgr}	28	367	9.225	9.084	10.996
poss. χ^2	35	1348	8.347	8.152	10.222
mut. spec.	33	666	8.332	8.182	10.390

Table 12. Results of possibilistic network learning with local structure (unrestricted).

to base selecting another parent on the score for a full decision tree, and to use local structure learning only to simplify this tree afterwards.

Bibliography

S.K. Andersen, K.G. Olesen, F.V. Jensen, and F. Jensen. HUGIN — A shell for building Bayesian belief universes for expert systems. *Proc. 11th Int. J. Conf. on Artificial Intelligence*, 1080–1085, 1989

P.W. Baim. A Method for Attribute Selection in Inductive Learning Systems. *IEEE Trans. on Pattern Analysis and Machine Intelligence*, 10:888-896, 1988

C. Borgelt and R. Kruse. Evaluation Measures for Learning Probabilistic and Possibilistic Networks. *Proc. 6th IEEE Int. Conf. on Fuzzy Systems (FUZZ-IEEE'97)*, Vol. 2:pp. 1034–1038, Barcelona, Spain, 1997

C. Borgelt and R. Kruse. Some Experimental Results on Learning Probabilistic and

eval. measure	num. of conds.	num. of params.	network quality		
			avg.	min.	max.
S_{gain}	34	752	8.584	8.349	10.500
S_{gr}	21	215	9.637	9.450	11.254
S_{sgr}	28	361	9.252	9.110	11.008
poss. χ^2	35	1347	8.348	8.152	10.222
mut. spec.	33	674	8.332	8.182	10.390

Table 13. Results of possibilistic network learning with local structure (levelwise).

eval. measure	num. of conds.	num. of params.	network quality		
			avg.	min.	max.
S_{gain}	31	1566	8.678	8.566	10.404
S_{gr}	18	182	9.627	9.446	11.202
S_{sgr}	28	455	9.074	8.948	10.812
poss. χ^2	35	1349	8.348	8.162	10.224
mut. spec.	33	621	8.402	8.262	10.502

Table 14. Results of possibilistic network learning with local structure preserving the global structure.

Possibilistic Networks with Different Evaluation Measures. *Proc. 1st Int. Joint Conference on Qualitative and Quantitative Practical Reasoning (ECSQARU/FAPR'97)*, pp. 71–85, Springer, Berlin, Germany, 1997)

C. Borgelt and R. Kruse. *Graphical Models — Methods for Data Analysis and Mining.* J. Wiley & Sons, Chichester, United Kingdom 2002

C. Boutilier, N. Friedman, M. Goldszmidt, and D. Koller. Context Specific Independence in Bayesian Networks. *Proc. 12th Conf. on Uncertainty in Artificial Intelligence (UAI'96)*, Portland, OR, 1996

L. Breiman, J.H. Friedman, R.A. Olshen, and C.J. Stone. *Classification and Regression Trees*, Wadsworth International Group, Belmont, CA, 1984

W. Buntine. Theory Refinement on Bayesian Networks. *Proc. 7th Conf. on Uncertainty in Artificial Intelligence*, pp. 52–60, Morgan Kaufman, Los Angeles, CA, 1991

D.M. Chickering, D. Heckerman, and C. Meek. A Bayesian Approach to Learning Bayesian Networks with Local Structure. *Proc. 13th Conf. on Uncertainty in Artificial Intelligence (UAI'97)*, pp. 80–89, Morgan Kaufman, San Franscisco, CA, 1997

C.K. Chow and C.N. Liu. Approximating Discrete Probability Distributions with Dependence Trees. *IEEE Trans. on Information Theory* 14(3):462–467, IEEE 1968

G.F. Cooper and E. Herskovits. A Bayesian Method for the Induction of Probabilistic Networks from Data. *Machine Learning* 9:309–347, Kluwer 1992

J. Gebhardt and R. Kruse. The context model — an integrating view of vagueness and uncertainty *Int. Journal of Approximate Reasoning* 9:283–314, 1993

J. Gebhardt and R. Kruse. POSSINFER — A Software Tool for Possibilistic Inference. In: D. Dubois, H. Prade, and R. Yager, eds. *Fuzzy Set Methods in Information Engineering: A Guided Tour of Applications*, Wiley 1995

J. Gebhardt and R. Kruse. Learning Possibilistic Networks from Data. *Proc. 5th Int. Workshop on Artificial Intelligence and Statistics*, 233–244, Fort Lauderdale, 1995

J. Gebhardt and R. Kruse. Tightest Hypertree Decompositions of Multivariate Possibility Distributions. *Proc. Int. Conf. on Information Processing and Management of Uncertainty in Knowledge-based Systems*, 1996

J. Gebhardt. *Learning from Data: Possibilistic Graphical Models.* Habil. thesis, University of Braunschweig, Germany 1997

D. Geiger and D. Heckerman. Advances in Probabilistic Reasoning. *Proc. 7th Conf. on Uncertainty in Artificial Intelligence (UAI'91)*, pp. 118–126, Morgan Kaufman, San Franscisco, CA, 1997

D. Heckerman. *Probabilistic Similarity Networks.* MIT Press 1991

D. Heckerman, D. Geiger, and D.M. Chickering. Learning Bayesian Networks: The Combination of Knowledge and Statistical Data. *Machine Learning* 20:197–243, Kluwer 1995

M. Higashi and G.J. Klir. Measures of Uncertainty and Information based on Possibility Distributions. *Int. Journal of General Systems* 9:43–58, 1982

K. Kira and L. Rendell. A Practical Approach to Feature Selection. *Proc. 9th Int. Conf. on Machine Learning (ICML'92)*, pp. 250–256, Morgan Kaufman, San Franscisco, CA, 1992

G.J. Klir and M. Mariano. On the Uniqueness of a Possibility Measure of Uncertainty and Information. *Fuzzy Sets and Systems* 24:141–160, 1987

I. Kononenko. Estimating Attributes: Analysis and Extensions of RELIEF. *Proc. 7th Europ. Conf. on Machine Learning (ECML'94)*, Springer, New York, NY, 1994

I. Kononenko. On Biases in Estimating Multi-Valued Attributes. *Proc. 1st Int. Conf. on Knowledge Discovery and Data Mining*, 1034–1040, Montreal, 1995

R.E. Krichevsky and V.K. Trofimov. The Performance of Universal Coding. *IEEE Trans. on Information Theory*, IT-27(2):199–207, 1983

R. Kruse, E. Schwecke, and J. Heinsohn. *Uncertainty and Vagueness in Knowledge-based Systems: Numerical Methods.* Series: Artificial Intelligence, Springer, Berlin 1991

R. Kruse, J. Gebhardt, and F. Klawonn. *Foundations of Fuzzy Systems*, John Wiley & Sons, Chichester, England 1994

S. Kullback and R.A. Leibler. On Information and Sufficiency. *Ann. Math. Statistics* 22:79–86, 1951

S.L. Lauritzen and D.J. Spiegelhalter. Local Computations with Probabilities on Graphical Structures and Their Application to Expert Systems. *Journal of the Royal Statistical Society, Series B*, 2(50):157–224, 1988

R. Lopez de Mantaras. A Distance-based Attribute Selection Measure for Decision Tree Induction. *Machine Learning* 6:81–92, Kluwer 1991

H.T. Nguyen. Using Random Sets. *Information Science* 34:265–274, 1984

J. Pearl. *Probabilistic Reasoning in Intelligent Systems: Networks of Plausible Inference (2nd edition).* Morgan Kaufman, New York 1992

D. Poole. Probabilistic Horn Abduction and Bayesian Networks. *Artificial Intelligence*, 64(1):81-129, 1993

J.R. Quinlan. *C4.5: Programs for Machine Learning*, Morgan Kaufman, 1993

L.K. Rasmussen. *Blood Group Determination of Danish Jersey Cattle in the F-blood Group System*. Dina Research Report no. 8, 1992

J. Rissanen. Stochastic Complexity. *Journal of the Royal Statistical Society (Series B)*, 49:223-239, 1987

A. Saffiotti and E. Umkehrer. PULCINELLA: A General Tool for Propagating Uncertainty in Valuation Networks. *Proc. 7th Conf. on Uncertainty in AI*, 323–331, San Mateo 1991

G. Shafer and P.P. Shenoy. Local Computations in Hypertrees. Working Paper 201, School of Business, University of Kansas, Lawrence 1988

P.P. Shenoy. Valuation-based Systems: A Framework for Managing Uncertainty in Expert Systems. Working Paper 226, School of Business, University of Kansas, Lawrence, 1991

J.E. Smith, S. Holtzman, and J.E. Matheson. Structuring Conditional Relationships in Influence Diagrams. *Operations Research*, 41(2):280–297, 1993

L. Wehenkel. On Uncertainty Measures Used for Decision Tree Induction. *Proc. IPMU*, 1996

X. Zhou and T.S. Dillon. A statistical-heuristic Feature Selection Criterion for Decision Tree Induction. *IEEE Trans. on Pattern Analysis and Machine Intelligence*, 13:834–841, 1991

SESSION III

APPLICATION of PLANNING
and DECISION MAKING

Coordination of the Supply Chain as a as a Distributed Decision Making Problem

Christoph Schneeweiss

Chair of Operations Research, University of Mannheim, Germany

Abstract The paper gives an overview on various kinds of coordination within the supply chain. In doing so, it relies on a taxonomy of distributed decision making describing the parties of a supply chain in a descending degree of connectedness. We are identifying those distributed decision making problems that appear to be most appropriate to capture the main features of a supply chain.

1 Introduction

A supply chain may be characterized as a logistic network of partially autonomous decision makers. Different segments of these networks are communicating with each other through flows of material and information, being controlled and coordinated by the activities of supply chain management (SCM). Obviously, since more than one decision maker is involved one has a typical distributed decision making (DDM) situation (Schneeweiss (1999)). Indeed, DDM has to do with the segregation of complex (often) multi-person decision problems into more tractable subsystems, and is subsequently concerned with coordinating these subsystems in accordance with an overall welfare function or in line with some equilibrium condition.

DDM comprises a large number of concepts and approaches which mainly differ in the information status of the supply chain partners and their mutual team-oriented or antagonistic behavior. Hence, focusing on the DDM character, it is interesting to analyze SCM as to the various kinds of DDM problems and approaches one encounters. In particular, it might be rewarding to identify those DDM problems that are most relevant for supply chain management.

In approaching this question let us first, in Section 2, characterize typical DDM problems in SCM. We then, in Section 3, give an overview over different problem classes in DDM and classify SCM problems according to these classes (Sec. 4). Subsequently, Section 5 identifies those DDM problems that seem to be most appropriate for SCM. Finally, Section 6 provides some concluding remarks.

2 DDM Problems in Supply Chain Management

In abstract terms, DDM problems arise because of two reasons:

1. The multi-level feature of design decisions, and
2. the coordination problems within the supply network.

As for any other complex management activity one may differentiate between the strategic, tactical, and operational level.

(1) The *strategic level* is mainly concerned with the design of the supply network. This involves problems of product design and the long-term selection of possible suppliers and customers.

(2) On the *tactical level* one is primarily interested in long- (and medium-) term investments, in the writing of long-term contracts, and in the design of market places.

(3) For the *operational level* one differentiates between a medium-term and a short-term level:

On the *medium-term level* (e.g., one or two years) one has the well-known network planning problems of determining medium-term production quantities and of adapting capacities.

On the *short-term level* the actual flow of material and information is scheduled and, in particular, one has to do with the (short-term) design of auctions.

Clearly, between and within these levels DDM problems may occur. Typical of the supply chain, however, is the *coordination* of different autonomous partners at all these levels, which traditionally is not treated in classical logistics.

To realize the importance and complexity of DDM problems in SCM consider the design of a medium-term contract between a supplier and a producer. Take the example of a contract that offers the supplier a bonus for a JIT delivery (Schneeweiss, Zimmer, and Zimmermann (2002)). This contract has as a direct consequence a more or less correct delivery. Simultaneously, however, to enable the supplier to deliver correctly, in most cases capacities must be adjusted. Thus, on the operational level one has not only to consider the short-term level of production scheduling but the medium-term level of capacity adaptation as well. Altogether, one has three interwoven DDM problems

1. One has the DDM problem of the (tactical) contract level and the operational level.
2. Within the operational level there is the DDM problem between the medium-term and the short-term decisions, and
3. one has the DDM problems between the producer and the supplier.

Fig. 1 illustrates these three DDM systems. In fact, the two vertical hierarchies (2) between the medium-term and short-term planning of the producer and the supplier are well known in PPC systems. New, however, are the DDM systems (3) between producer and supplier. On the medium-term level the producer informs the supplier of the capacity possibly needed while at the short-term level the producer specifies actual demand. In the mean time the supplier may possibly adapt (through his vertical hierarchy) his capacity so that short-term demand can be fulfilled. The three DDM systems will usually be of a distinctly different nature. Let us therefore give a brief overview and taxonomy of DDM systems.

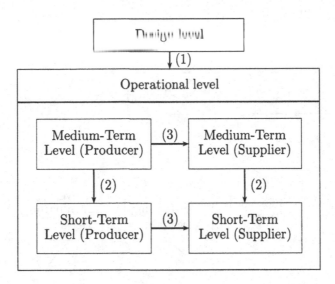

Figure 1. Three interwoven DDM systems in SCM

3 Some Basic Properties of DDM Systems

3.1   Classification of DDM Systems

A classification of DDM systems may be performed according to the number of decision makers involved, the symmetry or asymmetry of information, the team or non-team character, and the number of decisions that are communicated. Fig. 2 provides such a classification.

DDM systems can be divided into those involving only one decision making unit (DMU) and those which have to do with more than one DMU (see Fig. 2). The one-party setting leads, per definition, to conflict-free planning situations. For the multi-party case, on the other hand, one has to distinguish between team and non-team based decision situations. Apart from communication aspects, *team based DDM systems* are like one-party systems and are altogether denoted as *conflict-free DDM* problems.

For the *non-team based case*, the levels follow competitive goals in a self-centered way. This is typical for game theoretic settings and, in particular, for oligopolistic (competitive) markets, for coordinative (antagonistic) partners, and for *principal agent settings* (PA theory). For this latter theory the communication aspect plays a predominant role. The levels possess private information and are behaving opportunistically in that moral hazard and cheating has to be taken into account and the compromise is reached at the price of offering incentives. Antagonistic 'coordinative hierarchical DDM systems', on the other hand, may be viewed as PA systems having no private information or in which such

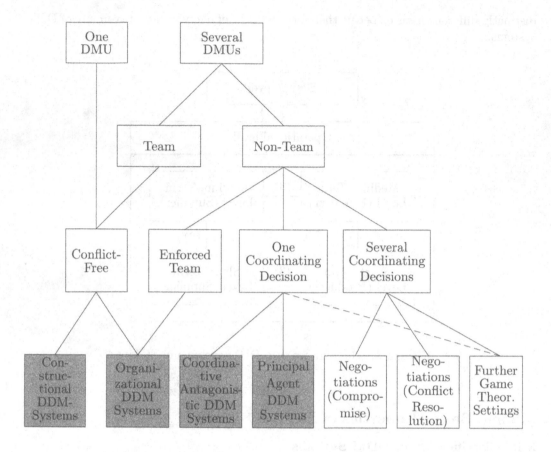

Figure 2. Classification of DDM systems

an information is not exploited opportunistically. Note that coordinative, competetive, and principal agent systems describe (one-shot) Stackelberg games.

The DDM systems at the bottom-line of Fig. 2 are describing settings of no more than one or two DMUs. This might not be sufficient, e.g., for multi-agent systems (MAS). The same holds for auction theory for which, as in PA theory, information asymmetry plays a predominant role and usually more than two parties are involved. Even for binary (complete) contracts a third party generally is necessary representing the court to supervise the contract.

In what follows, let us primarily concentrate on the four shaded DDM systems of Fig. 3 which are all of a hierarchical (Stackelberg) character and may again briefly be characterized as follows:

(1) Constructional DDM systems: team and information symmetry,

(2) Organizational DDM systems: team and information asymmetry,

(3) Coordinative DDM systems: non-team and information symmetry,

(4) Principal agent DDM systems: non-team and information asymmetry.

All four types of setting involve just one (important) decision communicated to one of the levels as opposed to negotiation situations depicted in one of the non-shaded boxes in the bottom of Fig. 2. Hence, the involved levels exhibit some non-symmetric (hierarchical) character (a so-called one-sided rationality) which, in game theory or oligopoly theory, is also known as Stackelberg property (e.g., see Varian (1992)). To be specific, let us therefore call the more dependent level 'base-level' and the independent one 'top-level'. Moreover, we are distinguishing between two cases of information asymmetry: For *weak information asymmetry* information is revealed when time passes on, *strict information asymmetry*, on the other hand, involves always two parties and might never be eliminated.

After this general classification let us now investigate coordination procedures in DDM systems. This will then help us to give a more formal characterisation of the different DDM directions and, in particular, of the important aspects of information and communication and of a team and non-team behavior of the involved decision makers in a supply chain.

3.2   Coordination of a Two-Level DDM System

A coordination scheme for DDM systems particularly described by the shaded boxes in Fig. 2 may simply be gained by the procedure indicated in Fig. 3.

The top-level, described by its model $M^T(C^T, A^T)$ with criterion C^T and action space A^T, is anticipating the base-model $M^B(C^B, A^B)$ in taking estimates $\hat{C}^B$ and $\hat{A}^B$, respectively. The optimal reaction of the anticipated base-level w.r.t. an instruction $IN(a^T)$, $a^T \in A^T$, is called anticipation function $AF(IN)$. Hence, the coupling of the two levels is given by the following set of functional equations

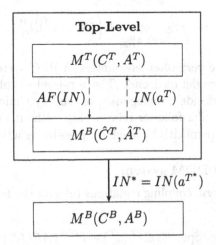

Figure 3. Structure of a hierarchical DDM

$$a^{T^*} = \arg \operatorname*{opt}_{a^T \in A^T} E\left\{ C^T \left[C^{TT}(a^T), C^{TB}\left(AF(IN(a^T)) \right) \right] | I_{t_0}^T \right\} \qquad (3.1\text{a})$$

$$AF(IN) \equiv \hat{a}^{B^*}(IN) = \arg \operatorname*{opt}_{\hat{a}^B \in \hat{A}_{IN}^B} E\left\{ \hat{C}_{IN}^B(\hat{a}^B) | \hat{I}_{IN}^B \right\} \qquad (3.1\text{b})$$

$$a^{B^*} = \arg \operatorname*{opt}_{a^B \in A_{IN^*}^B} E\left\{ C_{IN^*}^B(a^B) | I_{IN^*,t_1}^B \right\}. \qquad (3.1\text{c})$$

Note that the top-criterion consists of a 'top-down criterion' C^{TB} which explicitly depends on the anticipation function $AF(IN)$ and a 'private criterion' C^{TT} incorporating all remaining components. The anticipation function is taken to be the optimal decision of the anticipated base-model (Eq. 3.1b) as a function of IN, and the actual base-decision in $t_1 \geq t_0$ is given by Eq. 3.1c. The instruction IN is assumed to influence criterion and/or action space of the base-model and possibly even the information state. In particular, $\hat{I}_{IN}^B$ is the information the top-level assumes in t_0 the base-level might possess in t_1.

Let us now consider four cases which are of particular importance in coordinating the supply chain:

(1) top-down DDM systems,
(2) tactical-operational DDM systems,
(3) coordinative DDM systems, and
(4) PA systems.

(1) Top-down DDM system

For the top-down DDM system the coupling equations reduce to

$$a^{B^*} = \arg \operatorname*{opt}_{a^B \in A_{IN^*}^B} E\left\{ C_{IN^*}^B(a^B) | I_{IN^*,t_1}^B \right\} \qquad (3.2\text{a})$$

$$a^{B^*} = \arg \operatorname*{opt}_{a^B \in A_{IN^*}^B} E\left\{ C_{IN^*}^B(a^B) | I_{t_1}^B \right\}. \qquad (3.2\text{b})$$

This is the typical case one very often encounters in PPC systems in 'coordinating' an aggregate level with a more detailed one. The top-level simply sets the frame within which the base-level must decide. Analogously, for supply chain partners, the top-down system describes a strict leader-follower relationship without any reactive anticipation, i.e., the top-level does not (explicitly) consider a possible reaction as to its instruction.

(2) Tactical-operational DDM system

For this system the general coupling equations take on the form (Schneeweiss (1999), Chap. 4)

$$a^{T^*} = \arg \operatorname*{opt}_{a^T \in A^T} E\{ C^{TT}(a^T) + C^{TB}\left(AF(IN(a^T)) \right) | I_{t_0}^T \} \qquad (3.3\text{a})$$

$$\hat{a}^{B^*} = AF(IN) = \arg \operatorname*{opt}_{\hat{a}^B \in \hat{A}_{IN}^B} E\{ \hat{C}^B(\hat{a}^B) | \hat{I}^B \} \qquad (3.3\text{b})$$

$$a^{B^*} \quad = \quad \arg \; \underset{a^B \in A^B_{IN*}}{\mathrm{opt}} \; E\{\, C^B(a^B)|I^B_{t_1}\}. \tag{3.3c}$$

This type of DDM system is often used to describe the situation within a supply chain. It describes a team and the top-level's criterion is evaluating the entire supply chain. For a team situation C^{TB} and C^B are complementary, i.e., the top-level does not change the ranking of the base-decisions. Formally, C^{TB} is a monotonic function of $\hat{C}^B$ or even, in case of a channel coordination, $C^{TB} \equiv \hat{C}^B$. Moreover, $I^T_{t_0}$ and $I^B_{t_1}$ often describe a case of weak information asymmetry. Clearly, in a team setting the question of how profit is to be divided between the parties is of no relevance. Note, however, that there still exists some autonomy which is due to the separate existence of two different decision spaces A^T and A^B (or $\hat{A}^B$). The prominent case of a 'channel coordination' in SCM does belong to this class of problems.

(3) Deterministic coordinative systems

Deterministic coordinative (antagonistic) DDM systems may be described by the following

$$a^{T^*} = \arg \; \underset{a^T \in A^T_{AF}}{\mathrm{opt}} \; C^T(a^T, AF(IN)) \quad (t = t_0) \tag{3.4a}$$

$$IN = IN(a^T)$$

$$\hat{a}^{B^*} = \arg \; \underset{\hat{a}^B \in A^B_{IN}}{\mathrm{opt}} \; C^B_{IN}(\hat{a}^B) \quad (t = t_0) \tag{3.4b}$$

$$AF = AF(IN) = \hat{a}^{B^*}(IN)$$

$$a^{B^*} = \hat{a}^{B^*}(IN^*) \quad (t = t_1) \tag{3.4c}$$

They describe an antagonistic game-theoretic setting of the Stackelberg type. There is no information asymmetry involved and hence one has not to account for a possible opportunistic behavior resulting in moral hazard. In contrast to the tactical-operational DDM system there is no overall 'channel optimization' as expressed in Eq. 3.3a but solely a coordination achieved through a particular coordinating contract IN^* being determined by the Nash-equilibrium solution(s) of Eqs. 3.4. In particular cases, however, it proves to be possible to choose a contract such that a channel coordination may, in fact, be achieved. Note that *competetive DDM systems* (depicted in Fig. 2) are not coordinated by formal contracts but merely via a market competition resulting in a specific strategic behavior.

(4) Principal agent system

Principal agent (PA) theory again turns out to be a special case of the general coupling equations 3.1. Let us restrict ourselves to the hidden action case.

Defining $a^T = \phi$ to be the principal's incentive, a^B the agent's effort, u^T and u^B their respective utility functions, the standard model of PA theory may be written as

$$\phi^* = \arg \; \underset{\phi \in A^T}{\max} \; E\{u^T[P(AF(\phi)) - \phi(P(AF(\phi)))]|I^T_{t_0}\} \tag{3.5a}$$

$$AF(\phi) = \arg \max_{\hat{a}^B \in A_\phi^B} E\{u^B[\phi(P(\hat{a}^B)), \hat{a}^B]|I^B\}. \qquad (3.5b)$$

The incentive is optimizing the principal's (net)profit $P(AF)$ subject to the fact that the agent is optimizing his expected utility. Note that a profit-dependent incentive $\phi(P)$ has only to be paid because the principal is not assumed to be able to control the agent's action (strict information asymmetry: hidden action case, (e.g., see Schneeweiss (1999), Chap. 5)). In this antagonistic case both parties are trying to optimize their own criterion, i.e., there is, as in the 'coordinative case' (3), no primary intent to achieve 'channel optimization'.

4 The Nature of DDM Problems in Supply Chain Management

In characterizing DDM problems in SCM, let us follow the general idea of decreasing connectedness of the supply chain, starting with rather closely connected partners and ending up with fairly loosely related decision making units (see Fig. 4).

(1) First let us consider *constructional DDM problems*. These problems with their team character and symmetric state of information are typical for situations in classical logistics. Modern software in SCM is mostly of this type, i.e., one has still the situation of a single company in which it is in principle possible to exchange all relevant data (e.g., see RHYTHM (Trade Matrix) of i2 (RHYTHM (1999)), APO of SAP (SAP-APO (2000), and Stadtler and Kilger (2000)). That is, a complex decision problem is separated into local problems being centrally coordinated by a chief authority which, in principle, has access to all information.

(2) A slightly less tight connection is described by the organisational DDM system which is characterized by a team (or enforced team) situation and by *weak information asymmetry*, i.e., one has the typical tactical-operational type of system. Or, to put it differently, one has the connection of longer-term with short-term levels where the lower level's uncertainty is revealed (or at least reduced) at the time when a decision of this level has to be made. Again we have a SC situation which can be described within classical logistics.

(3) A proper supply chain setting might be characterized by partners having some private information (defining a state of strict information asymmetry) but who are still forming a team, i.e., the top-level is adopting the base-criterion and is achieving a channel optimization depending, of course, on the restrictions imposed by the base-level. The parties are following their own autonomous goals but they are supporting each other and are not exploiting their private information in an opportunistic way. In particular, profit sharing is not an issue. For a contract following these lines, e.g., see Schneeweiss and Zimmer (2002).

(4) As a next step one might consider supply chain contracts of partners that are symmetrically informed (forming a coordinative DDM system) or, in case of asymmetric information do abstain from opportunistic behavior. Like the previous case this type of coordination seems to be particularly adequate for a supply chain since the assumption of cheating is not very realistic for a longer term relationship. It

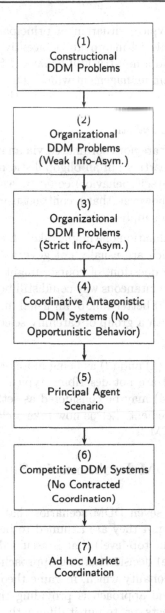

Figure 4. Grades of Connectedness within the Supply Chain

should be clear, however, that the partners in this setting are not interested in a channel optimization but merely in a 'partner-centered' coordination, i.e., in a contract being the result of a Nash equilibrium of the corresponding coupling equations.

(5) As a fifth step we define a coordination of asymmetrically informed partners which, in principle, are seeking for a cooperation but are behaving opportunistically. They do not have in mind the supply chain as a whole but are optimizing their own

local goals. This is the typical situation of principal agent systems in which a coordination is mainly achieved in providing incentives based on contracts being constructed mainly to avoid a non-truthful behavior. Such a setting turns out not to be typical of supply chain management which is usually looking for a long-term relationship and is thus basing its contracts on truthful behavior. Only for short term (purchasing) contracts which are not intended to be repeated, PA theory-based contracts seem to be realistic.

(6) Competitive DDM systems are not coordinated via an explicit contract but are just influencing each other as partners in an oligopolistic market competition. Such a coordination (through strategic behavior) could be as strict as a formal contract. One should have in mind, however, that a contract usually describes a more stable situation being typical of a supply chain.

(7) As the loosest kind of coordination one might take (short-term) ad hoc negotiations and market exchanges which are usually not governed by a known or prescribed coordination scheme. The question, of course, might be raised whether partners being coordinated in this spontaneous way could still be viewed as forming a supply chain or whether one should better talk of an usual market exchange. If, however, an auction is performed with a group of carefully selected partners, it loses its ad hoc character.

Summarizing, one could identify (1) and (2) as being more or less problems of traditional logistics, while (5), (6), and (7) are not describing typical (longer-term) supply chain coordinations. Hence, (3) and (4) may be considered as settings that are proper DDM problems of supply chain management. Let us now take a closer look at the significance of the above classification for SCM.

5 Identifying Proper DDM Problems in Supply Chain Management

Thus far the analysis identified seven DDM scenarios that may occur in SCM. Apart from scenarios (1) and (7) of Fig. 4 they are assumed to be primarily of a hierarchical nature. I.e., there is one level (the top-level) that is assumed to be the more active one. This implies that the (normative) decision making approach is 'located' in the top-level adopting some 'asymmetric' rationality which, in game theory, is known as Stackelberg property. Hence, the hierarchical approach is providing the top-decision maker with suggestions for reasonable contracts, or, to put it differently, the analysis presented thus far might be the starting point for actually negotiating a contract.

Having this asymmetric perspective in mind we may now identify contracts that are of particular interest in SCM. These contracts rely on scenarios (3) and (4) in Fig. 4 which can be characterized as describing supply chain partners that do not behave opportunistically, i.e., they are in a team (case (3) in Fig. 4, channel coordination) or in a non-team setting (case (4) in Fig. 4, partner-centered (game-theoretic) coordination). This is because, for a long-term relationship, cheating and non-truthful opportunistic behavior would not remain secret, and hence seems not to be a reasonable basis for a (long-lasting) supply contract. Hence, for SC contracts one usually has not the typical setting of contracts based on principal agent theory. However, considering case (3) of Fig. 4 the partners are not likely to disclose all their (relevant) private knowledge. Hence, in this situation, just as in principal agent relationships, incentives are reasonable. These incentives, though, are not used to counteract moral hazard or to allocate profits to the parties involved but they are often utilized to inform the other party about the consequences his action has for the overall performance (e.g., see Schneeweiss and Zimmer (2002)). Accordingly, not the single partners are optimizing their own profit but it is the supply chain as a whole that is optimized, and hence it might be more appropriate to talk of an 'agreement' rather than of a 'contract'. Clearly, in some cases, it might be reasonable that in case of such a 'channel coordination' an agreement should be augmented by some profit-sharing regulations. On the other hand, regarding the antagonistic coordination setting of case (4) in Fig. 4 a coordination can only be achieved in finding an equilibrium which does not necessarily imply a 'channel optimization', i.e., we just have a partner-centered coordination. For short-term contracts (e.g., one-time purchasing contracts and contracts that are not intended to be continued or repeated), PA models should be employed in case of private knowledge.

Reading reversely, Fig. 4 could be interpreted as describing the possible development of a coordination of a supply chain. The 'market scenarios' (7) and (6) describe a search for SC partners which might result in a first general contract. This contract is then refined leading in a first step to a partner-centered coordination and later, after a sufficient time of truthful cooperation has elapsed, might possibly result in a team-oriented channel coordination. Clearly, this will only occur for a win-win situation and for an allocation of the additional profit which is acceptable for all partners. As a last step this highly reliable cooperation might even end up in a centralistic coordination described by scenarios (2) and (1). Up to now this highly important and interesting transition of the nature of contracts (which one often meets in practice) has not yet found its counterpart in theory. Let us therefore reflect a little longer on this crucial kind of problem.

Looking at SCM in more general terms, one encounters expressions like collaboration, coordination, cooperation, or integration. They all describe some relationship between the partners of the supply chain. Though these terms are generally not used in a uniform way, let us employ them here to give a more comprehensive and detailed description of connectedness within the supply chain. Let us use the term 'coordination' to describe the decision-analytic aspects of a relationship. Collaboration, on the other hand, describes the more general aspects of a relationship. Collaborating partners might take various kinds of actions which might not be describable within a decision-analytic framework. In particular, collaboration might lead to a transformational (intrinsic) change of the partner's preferences and to a mutual adaptation of the respective decision spaces. Thus it is just the 'collaborative environment' that might promote the transition of a loose connectedness to a closer relationship.

As we described in Fig. 4, one might consider a transition from a PA situation, via a cooperative game-theoretic coordination, to a channel coordination. During this transition, the partners are gaining a more faithful relationship. They first abandon a non-truthful behavior and, in a second step, they do no longer insist on obtaining a certain share of the overall profit of a SC. From a more theoretic point of view, it is important to realize that this transition from the non-team to the team situation can entirely be described by the coupling equations, i.e., it is not necessary to change the mathematical formalism. Often one has (more or less) the same formalism, and it is only a matter of interpretation that differentiates a team from a cooperative game-theoretic coordination. In fact, for antagonistic partners, in coordinating, real side payments are achieving the coordination, i.e., the cooperation. In a team one has coordinating side payments as well, but in this case the payments cancel (i.e., the base-level is paying the amount the top-level is losing) and are only used as coordinating signals (see the final discussion in Schneeweiss and Zimmer (2002)).

Between the main stages (4) and (3), numerous intermediate stages could be considered. As a necessary condition for a team we postulated, in Section 3.2, C^{TB} to be complementary to $\hat{C}^B$, or in more mathematical terms, C^{TB} must be a monotonic function of $\hat{C}^B$. In addition, adopting an additive top-criterion we realized that channel coordination belongs to the class of tactical-operational DDM systems. A channel coordination (i.e., scenario (3) in Fig. 4) still allows for some private information and for some (aspiration) level of income being retained for each party involved in a SC. Moreover, rather than simply adding C^{TT} and C^{TB} (with $C^{TB} = \hat{C}^B$), one would still have a team-like attitude if one took for the top-criterion a convex combination of the top-level's and base-level's profit: $C^T :=$ 'α (profit of top-level) $+(1 - \alpha)$ (profit of base-level)'. Excluding the extreme cases $\alpha = 1$ (i.e., game-theoretic coordination) and $\alpha = 0$ (i.e., altruistic behavior) one could generate, in changing $\alpha \in (0, 1)$, numerous grades of such a behavior.

The transition from (5) to (3) and even further to (2) and (1) would usually involve some collaborative behavior. If the partners are building a firm team, they would try to employ all integration measures available. This might particularly result in deliberately changing the Stackelberg leadership position in gaining a better channel optimum. More generally, collaboration might result in mutually employing a suitable optimization procedure to solve the coupling equations, or, even more generally, as mentioned before, one might alter the decision spaces. In this light, the logistics-oriented supply chain optimizations (like the software packages mentioned in Sec. 4 (1)) can be considered as optimization procedures within a contract of a stable team.

6 Summary and Concluding Remarks

Design decisions and coordinations have been identified as the two main reasons for DDM in SCM. While design decisions are not specific for SCM, we recognized coordinations as the main challenge for SCM. We therefore classified various types of coordination schemes according to a general taxonomy developed in DDM.

We related the degree of connectedness within the supply chain to constructional and organizational DDM systems in case of a team (or enforced team) and to principal agent and further game theoretic DDM systems in case of a non-team relationship. Moreover, we related the approaches developed in different sciences to the DDM classification and consequently to the various degrees of connectedness within the supply chain.

As one of the main problems in DDM-oriented SCM we identified the construction of coordinating contracts (or agreements) within a team situation for partners having private knowledge and for antagonistic partners having no private knowledge or which do not behave opportunistically. In general, we adopted the perspective of asymmetric rationality (hierarchical and Stackelberg setting) which could be used as the starting point of a possible negotiation (or renegotiation). If, however, the one-shot (static) game is representing the contractual situation of a SC, one should incorporate a possible future behavior within the cooperation which means that, for a longer lasting relationship, non-truthful behavior should not enter the simple static description.

Generally, in capturing the contractual situation of a SC, the solution of the coupling equations can only be considered as a first step. Renegotiations are necessary, possibly describing the transition from a non-team to a team situation. In doing so, the role of a possible collaboration should be investigated and incorporated into the decision-analytic description. Clearly, it is not enough simply to describe a transition of the coupling equations, but it would be necessary to anticipate future behavior. The framework the coupling equations are providing seems to be rich enough to capture main features of these rather involved features of an extended contractual Supply Chain Management.

Bibliography

Ch. Schneeweiss. *Hierarchies in Distributed Decision Making*. Springer, Berlin, Heidelberg, New York, 1999.

Ch. Schneeweiss, K. Zimmer, and M. Zimmermann. *The Design of Contracts to Coordinate Operational Interdependencies within the Supply Chain*. Preprint of the 12th International Working Seminar on Production Economics. Igls, Feb. 18-22, 2002.

Ch. Schneeweiss and K. Zimmer. *Hierarchical Coordination Mechanism within the Supply Chain*. To be published in European Journal of Operations Research (EJOR), 2003.

RHYTHM. Supply Chain Software Package of i2. Http://www. tradematrix.com, 1999.

SAP-APO. Advanced Planner and Optimizer (APO). Http://www. sap.com, 2000.

H. Stadtler and Ch. Kilger, editors. *Supply Chain Management and Advanced Planning*. Springer, Berlin, Heidelberg, New York, 2000.

H. Varian. *Microeconomic Analysis*. 3. ed., New York, 1992.

Fuzzy Decision Theory
Intelligent Ways for Solving Real-World Decision Problems and for Solving Information Costs

Heinrich J. Rommelfanger

Institute of Statistics and Mathematics, Goethe-University Frankfurt am Main, Germany

Abstract: Looking at modern theories in management science and business administration, one recognizes that many of these conceptions are based on decision theory in the sense of von Neumann and Morgenstern. However, empirical surveys reveal that the normative decision theory is hardly used in practice to solve real-life problems. This neglect of recognized classical decision concepts may be caused by the fact that the information necessary for modeling a real decision problem is not available, or the cost for getting this information seems too high. Subsequently, decision makers (DM's) abstain from constructing decision models.

As the fuzzy set theory offers the possibility to model vague data as precise as a person can describes them, a lot of decision models with fuzzy components are proposed in literature since 1965. But in my opinion only fuzzy consequences and fuzzy probabilities are important for practical applications. Therefore, this paper is restricted to these subjects. It is shown that the decision models with fuzzy utilities or/and fuzzy probabilities are suitable for getting realistic models of real world decision situations. Moreover, we propose appropriate instruments for selecting the best alternative and for compiling a ranking of the alternatives. As fuzzy sets are not well ordered, this should be done in form of an interactive solution process, where additional information is gathered in correspondence with the requirements and under consideration of cost–benefit relations. This procedure leads to a reduction of information costs.

1   Classical Decision Model

Decision models in the sense of von Neumann and Morgenstern (1953) are well known in management science and a lot of modern theories in business administration are based on these conceptions. On the other side, empirical surveys reveal that the normative decision theory is hardly used in practice to solve real-life problems. This was discussed in detail by R.M. Cyert and J.G. March in their famous book "Behavioral theory of the firm" of 1963. This scientific discoveries were later underpined by empirical studies of Kivijärvi; Korhonen; Wallenius (1986); Lilien (1987); Tingley (1987); Meyer von Selhausen (1989); Fandel; Francois; Gulatz (1994), which came to the result that only few operations research methods are used in practice and that a lot of applications proposed in OR literature are not transformed into practical applications.

This neglect of recognized classical decision concepts may be caused by the fact that the information necessary for modeling a real decision problem is not available, or the cost for getting this information seems too high.

In order to design a decision problem by classical decision models, the decision maker (DM) must be able to specify the following elements:

1. A set A of actions, $A = \{a_1, a_2, \ldots, a_m\}$,

2. A set S of possible events, $S = \{s_1, s_2, \ldots, s_n\}$,

3. A result associated with each act-event combination,
   $g_{ij} = g(a_i, s_j)$, $i = 1, 2, \ldots, m$; $j = 1, 2, \ldots, n$. G is the set of possible values g_{ij}.

4. The degree of knowledge with regard to the chance of occurrence of each event. Usually it is assumed that the DM knows the probability distribution $p(s_j)$.

5. A criterion by which a course of action is selected:
   In literature, the Bernoulli-criterion is recommended, i.e. the expected utility should be maximized:

$$E(a^*) = \operatorname*{Max}_{a_i \in A} E(a_i) = \operatorname*{Max}_{a_i \in A} \sum_{j=1}^{n} u(g(a_i, s_j)) \cdot p(s_j) \tag{1}$$

6. A posteriori probability distribution:
   In classical decision models the only chance for getting a better solution is to use additional information $X = \{x_1, x_2, \ldots, x_K\}$. Knowing the likelihoods $p(x_k | s_j)$, the priori probability distribution $p(s_j)$ can be substituted by the posteriori probability distribution

$$p(s_j | x_k) = \frac{p(x_k | s_j) \cdot p(s_j)}{\sum_{j=1}^{n} p(x_k | s_j) \cdot p(s_j)} \quad \text{Bayes's formula} \tag{2}$$

With the additional information that x_k is observed the optimal action $a^*(x_k)$ satisfies the term

$$E(a^*(x_k)) = \operatorname*{Max}_{a_i \in A} \sum_{j=1}^{n} u(g(a_i, s_j)) \cdot p(s_j | x_k). \tag{3}$$

The expected value of additional information is

$$E(X) = \sum_{k=1}^{K} E(a^*(x_k)) \cdot p(x_k) - E(A^*). \tag{4}$$

Therefore, additional information should be used, if the cost for getting this information is smaller than E(X).

Nevertheless, in complex decision situation a DM needs a support by using a decision model based on a normative decision theory, but this must be done in a more realistic way. This can be done by adopting an adaptive solution process and taking into account the fact that human DM have only a restricted rational behavior. Essential parts of these improvements are

✓ Moving away from the classical optimization idea in favor of satisfying approaches.

✓ Realistic design of the search for alternative acts and results.

✓ Modeling vague data by fuzzy sets,]

✓ Abstention from expensive information

✓ Stopping raising of additional information for improving the structure and the data of the decision model by weighting benefits and costs.

Since the paper "Fuzzy Sets" of Lofti A. Zadeh was published in 1965, the fuzzy sets theory has been considered as a new way for modeling more realistic decision models. Especially between 1975 and 1985 several decision models with various fuzzy components were introduced. Without any claim on completeness, the following fuzzy elements have been proposed for use in decision models:

1. Fuzzy acts $\tilde{D}_h = \{(a_i, \mu_{D_h}(a_i)) \mid a_i \in A\}$, $h = 1, \ldots, H$, Tanaka, Oukuda; Asai (1976).

2. Fuzzy events $\tilde{Z}_r = \{(s_j, \mu_{Z_r}(s_j)) \mid s_j \in S\}$, $r = 1, \ldots, R$, Tanaka, Oukuda; Asai (1976).

3. Fuzzy probabilities $\tilde{P}_j = \tilde{P}(s_j) = \{(p, \mu_{P_j}(p) \mid p \in [0,1]\}$, Watson; Weiss; Donell (1979); Dubois; Prade (1982); Whalen (1984).

4. Fuzzy utility values $\tilde{U}_{ij} = \tilde{U}(a_i, s_j) = \{(u, \mu_{U_{ij}}(u) \mid u \in U\}$, where U is the set of given crisp utilities associated with each act-event combination, JAIN (1976); Watson; Weiss; Donell (1979); Yager (1979); Rommelfanger (1984); Whalen (1984).

5. Fuzzy information $\tilde{Y}_t = \{(x_k, \mu_{Y_t}(x_k) \mid x_k \in X\}$, Tanaka; Oukuda; Asai (1976); Sommer (1980).

6. Moreover, some authors propose to substitute the probability distribution $p(s_j)$ by a possibility distribution $\pi(s_j)$, see e.g. Yager (1979); Whalen (1984); Dubois and Prade (1988, 1998). They assume that utilities can be measured on an ordinal scale only and therefore expected values do not exist.

These new ideas, however were not applied to practice, either because they did not become known to the public or because they are of little use for real decision problems.

In my opinion the latter statement is correct as far as the points 1, 2, 5 and 6 are concerned:

• DMs need workable but not fuzzy acts.

• In real problems, the events and the information are usually described in a fuzzy way. In these cases one is able to assign directly probabilities to those elements; that means that we have probability distributions $p(\tilde{Z}_r)$ and $p(\tilde{Y}_t \mid \tilde{Z}_r)$ and values directly associated with the combinations $(a_i, \tilde{Z}_r)$, $i = 1, 2, \ldots, m$; $r = 1, 2, \ldots, R$. Therefore, we can use the classical procedure as we replace s_j by $\tilde{Z}_r$ and x_k by $\tilde{Y}_t$. But for simplifying the presentation, we will use crisp notations in this paper.

- In my opinion persons have no idea how to interpret possibility degrees in contrast to the interpretation of probability degrees. Moreover, possibility measures allow no addition or multiplication but only the comparison of possibility values by using the min- or max-operator. Therefore, I prefer to use probabilities, even though we have only subjective ones.

I think the best chance of increasing the acceptance of decision models in practice is to use fuzzy utilities and fuzzy probabilities. Therefore, this paper concentrates on these two extensions. At first, the use of fuzzy utilities or fuzzy values associated with each act-event combination is discussed. In this case the well known Bernoulli-principle can be extended to the fuzzy model. Moreover, if it is possible to get additional information of a test market, we can improve the solution by using a posteriori probability distributions. The concept of „value of additional information" can also be extended to fuzzy models by using fuzzy values of information, (see Rommelfanger 1994).

Crucial topics of decision models with fuzzy utilities are:
- The modeling of fuzzy utilities associated with each act-event combination,
- the definition of expected utility values,
- the preference orderings of expected utility values.

In addition to fuzzy utilities fuzzy probabilities will be used in the second part of the paper. There the main problem is the calculation of expected utility values.

2 Modeling Fuzzy Values Associated With Each Act-Event Combination

One of the most difficult problems in classical decision theory is the transformation of the consequences $g_{ij} = g(a_i, s_j)$ in utility values $u_{ij} = u(g(a_i, s_j))$. Working with fuzzy values $\tilde{G}_{ij} = \{(g, \mu_{\tilde{G}_{ij}}(g)) | g \in G)\}$, the classical methods of constructing utility functions can be transferred to fuzzy sets.

In this paper, the question how to get (fuzzy) utility functions is not discussed. Therefore, it is assumed that the DM knows the utility function $u = u(g_{ij})$; then the fuzzy results are mapped in the fuzzy utilities $\tilde{U}_{ij} = \{(u(g), \mu_{\tilde{G}_{ij}}(g)) | g \in G\}$ or alternatively the DM is able to specify directly utility values $\tilde{U}_{ij} = \{(u, \mu_{\tilde{U}_{ij}}(u)) | u \in U\}$, where U is the possible set of crisp utility values. Obviously in the case of risk neutrality, the fuzzy consequences $\tilde{G}_{ij}$ can be used instead of fuzzy utilities $\tilde{U}_{ij}$.

In literature the values $\tilde{G}_{ij}$ or $\tilde{U}_{ij}$ are usually modeled in form of triangular fuzzy numbers. But this shape with an unique culmination requires too much information. Therefore, the application of fuzzy intervals is often more realistic. On the other side, a DM can usually do better than just specify simple linear increasing and decreasing functions; often the DM can characterize the fuzzy interval in more detail. Concerning the membership function of fuzzy intervals it is almost impossible to describe these functions in detail because this requires immense investigations. Therefore, the membership functions used in practice are in general a rough description of the subjective imagination.

An efficient way of getting suitable membership functions is the following procedure. For a given membership degree, the α-level set or α-cut of a fuzzy set $\tilde{A}$ is defined as: $A_\alpha = \{x \in X | \mu_{\tilde{A}}(x) \geq \alpha\}$ where $\alpha \in [0,1]$. The importance of the α-cuts is obvious in the representation theorem; it says that a fuzzy set $\tilde{A}$ is completely characterized by the accompanying family of α-level-sets because the membership function of $\tilde{A}$ can be written as

$$\mu_{\tilde{A}}(x) = \sup\{\alpha \in [0,1] | x \in A_\alpha\} \tag{5}$$

Therefore, an approximation of a fuzzy set can be constructed by using few α-cuts.

At first the DM specifies some membership values and relates them to special meanings. This step can be clarified by using three levels, which appear to be sufficient for practical applications.

$\alpha = 1$: $\mu_{\tilde{U}_{ij}}(u) = 1$ means that u has the highest chance of belonging to the set of utility values associated with the act-event combination (a_i, s_j). In any case, values of u with $\mu_{\tilde{U}_{ij}}(u) = 1$ are primarily relevant for the decision.

$\alpha = \lambda$: $\mu_{\tilde{U}_{ij}}(u) \geq \lambda$ means that the decision maker is willing to accept u as an available value for the time being. A value u with $\mu_{\tilde{U}_{ij}}(u) \geq \lambda$ has a good chance of belonging to the set of utility values associated with the act-event combination (a_i, s_j). Corresponding values of u are relevant for the decision.

$\alpha = \varepsilon$: $\mu_{\tilde{U}_{ij}}(u) < \varepsilon$ means that u has only a very little chance of belonging to the set of utility values associated with the act-event combination (a_i, s_j). The DM is willing to neglect the values u with $\mu_{U_{ij}}(u) < \varepsilon$.

Accordingly the DM should specify numbers $\underline{u}_{ij}^1, \overline{u}_{ij}^1, \underline{u}_{ij}^\lambda, \overline{u}_{ij}^\lambda, \underline{u}_{ij}^\varepsilon, \overline{u}_{ij}^\varepsilon \in \mathbf{R}$, such that

$$\mu_{\tilde{U}_{ij}}(u) \begin{cases} =1 & \text{if } u \in [\underline{u}_{ij}^1, \overline{u}_{ij}^1] \\ <1 & \text{elsewhere} \end{cases} \qquad \mu_{\tilde{U}_{ij}}(u) \begin{cases} \geq \lambda & \text{if } u \in [\underline{u}_{ij}^\lambda, \overline{u}_{ij}^\lambda] \\ <\lambda & \text{elsewhere} \end{cases}$$

and $\quad \mu_{\tilde{U}_{ij}}(u) \begin{cases} \geq \varepsilon & \text{if } u \in [\underline{u}_{ij}^\varepsilon, \overline{u}_{ij}^\varepsilon] \\ <\varepsilon & \text{elsewhere} \end{cases}$

The lower the DM's information, the larger are the intervals $[\underline{u}_{ij}^{\alpha}, \overline{u}_{ij}^{\alpha}]$, $\alpha = 1$, λ, ε. The special case $\underline{u}_{ij}^1 = \overline{u}_{ij}^1$ can also be imagined, but it is rarely realistic to assume that all coefficients $\tilde{U}_{ij}$ are fuzzy numbers.

Consequently the polygon connecting $(\underline{u}_{ij}^{\varepsilon}, \varepsilon)$ over $(\underline{u}_{ij}^{\lambda}, \lambda)$, $(\underline{u}_{ij}^1, 1)$, $(\overline{u}_{ij}^1, 1)$, $(\overline{u}_{ij}^{\lambda}, \lambda)$ to $(\overline{u}_{ij}^{\varepsilon}, \varepsilon)$ is a suitable approximation of the membership function of $\tilde{U}_{ij}$ on $[\underline{u}_{ij}^{\varepsilon}, \overline{u}_{ij}^{\varepsilon}]$.

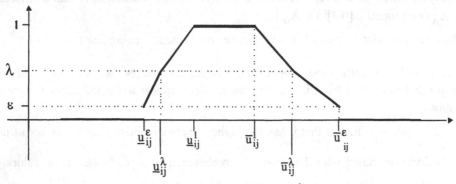

Figure 1. Membership function of $\tilde{U}_{ij} = (\underline{u}_{ij}^{\varepsilon}; \underline{u}_{ij}^{\lambda}; \underline{u}_{ij}^1; \overline{u}_{ij}^1; \overline{u}_{ij}^{\lambda}; \overline{u}_{ij}^{\varepsilon})^{\varepsilon,\lambda}$

A fuzzy interval $\tilde{U}_{ij}$ with this kind of membership function is characterized by $(\underline{u}_{ij}^{\varepsilon}; \underline{u}_{ij}^{\lambda}; \underline{u}_{ij}^1; \overline{u}_{ij}^1; \overline{u}_{ij}^{\lambda}; \overline{u}_{ij}^{\varepsilon})^{\varepsilon,\lambda}$. For simplification this special fuzzy set is called a fuzzy interval of the ε-λ-type. If the DM has more information, it is possible to specify additional membership levels and additional points $(u, \mu_{\tilde{U}_{ij}}(u))$ of the polygon line. The differences $\underline{u}_{ij}^1 - \underline{u}_{ij}^{\varepsilon}$ and $\overline{u}_{ij}^1 - \overline{u}_{ij}^{\varepsilon}$ are called spreads of the fuzzy interval $\tilde{U}_{ij}$; they present a measure of the fuzziness of $\tilde{U}_{ij}$.

It is an advantage of fuzzy intervals of the ε-λ-type that the arithmetic operations can be calculated extremely simple. In the following, the extended arithmetic operations are symbolized by $\oplus$ extended addition, $\ominus$ extended subtraction, $\otimes$ extended multiplication.

$$(\underline{a}^{\varepsilon}, \underline{a}^{\lambda}, \underline{a}^1, \overline{a}^1, \overline{a}^{\lambda}, \overline{a}^{\varepsilon})^{\varepsilon,\lambda} \oplus (\underline{b}^{\varepsilon}, \underline{b}^{\lambda}, \underline{b}^1, \overline{b}^1, \overline{b}^{\lambda}, \overline{b}^{\varepsilon})^{\varepsilon,\lambda}$$

$$= (\underline{a}^{\varepsilon} + \underline{b}^{\varepsilon}, \underline{a}^{\lambda} + \underline{b}^{\lambda}, \underline{a}^1 + \underline{b}^1, \overline{a}^1 + \overline{b}^1, \overline{a}^{\lambda} + \overline{b}^{\lambda}, \overline{a}^{\varepsilon} + \overline{b}^{\varepsilon})^{\varepsilon,\lambda} \tag{6}$$

$$(\underline{a}^{\varepsilon}, \underline{a}^{\lambda}, \underline{a}^1, \overline{a}^1, \overline{a}^{\lambda}, \overline{a}^{\varepsilon})^{\varepsilon,\lambda} \ominus (\underline{b}^{\varepsilon}, \underline{b}^{\lambda}, \underline{b}^1, \overline{b}^1, \overline{b}^{\lambda}, \overline{b}^{\varepsilon})^{\varepsilon,\lambda}$$

$$= (\underline{a}^{\varepsilon} - \overline{b}^{\varepsilon}, \underline{a}^{\lambda} - \overline{b}^{\lambda}, \underline{a}^1 - \overline{b}^1, \overline{a}^1 - \underline{b}^1, \overline{a}^{\lambda} - \underline{b}^{\lambda}, \overline{a}^{\varepsilon} - \underline{b}^{\varepsilon})^{\varepsilon,\lambda} \tag{7}$$

$$(\underline{a}^{\epsilon},\underline{a}^{\lambda},\underline{a}^{1},\overline{a}^{1},\overline{a}^{\lambda},\overline{a}^{\epsilon})^{\epsilon,\lambda} \otimes (\underline{b}^{\epsilon},\underline{b}^{\lambda},\underline{b}^{1},\overline{b}^{1},\overline{b}^{\lambda},\overline{b}^{\epsilon})^{\epsilon,\lambda}$$

$$= (\underline{a}^{\epsilon} \cdot \underline{b}^{\epsilon}, \underline{a}^{\lambda} \cdot \underline{b}^{\lambda}, \underline{a}^{1} \cdot \underline{b}^{1}, \overline{a}^{1} \cdot \overline{b}^{1}, \overline{a}^{\lambda} \cdot \overline{b}^{\lambda}, \overline{a}^{\epsilon} \cdot \overline{b}^{\epsilon})^{\epsilon,\lambda} \tag{8}$$

For the special case that $(\underline{b}^{\epsilon},\underline{b}^{\lambda},\underline{b}^{1},\overline{b}^{1},\overline{b}^{\lambda},\overline{b}^{\epsilon})^{\epsilon,\lambda}$ is a crisp number b $(b,b,b,b,b,b)^{\epsilon,\lambda}$ the term (8) can be simplified to

$$(\underline{a}^{\epsilon},\underline{a}^{\lambda},\underline{a}^{1},\overline{a}^{1},\overline{a}^{\lambda},\overline{a}^{\epsilon})^{\epsilon,\lambda} \otimes b = (\underline{a}^{\epsilon} \cdot b, \underline{a}^{\lambda} \cdot b, \underline{a}^{1} \cdot b, \overline{a}^{1} \cdot b, \overline{a}^{\lambda} \cdot b, \overline{a}^{\epsilon} \cdot b)^{\epsilon,\lambda} \tag{9}$$

3 Fuzzy Expected Values

As each real number a can be modeled as a fuzzy number

$$\tilde{A} = \{(x, \mu_{\tilde{A}}(x)) \mid x \in \mathbf{R}\} \quad \text{with} \quad \mu_{\tilde{A}}(x) = \begin{cases} 1 & \text{if} \quad x = a \\ 0 & \text{else} \end{cases}$$

we can assume that each act-event combination (a_i, s_j) is valued by a fuzzy interval

$$\tilde{U}_{ij} = (\underline{u}_{ij}^{\epsilon}; \underline{u}_{ij}^{\lambda}; \underline{u}_{ij}^{1}; \overline{u}_{ij}^{1}; \overline{u}_{ij}^{\lambda}; \overline{u}_{ij}^{\epsilon})^{\lambda,\epsilon}, \quad i = 1,...,m; j = 1,...,n.$$

Moreover, if the DM is able to specify a priori probabilities $p(s_j)$, $j = 1, 2,...,n$, the expected value of each alternative a_i can be calculated as:

$$\tilde{E}(a_i) = (\underline{E}_i^{\epsilon}; \underline{E}_i^{\lambda}; \underline{E}_i^{1}; \overline{E}_i^{1}; \overline{E}_i^{\lambda}; \overline{E}_i^{\epsilon})^{\epsilon,\lambda} = \tilde{U}_{i1} \otimes p(s_1) \oplus \cdots \oplus \tilde{U}_{in} \otimes p(s_n) \tag{10}$$

where $\underline{E}_i^{\alpha} = \sum_{j=1}^{n} \underline{u}_{ij}^{\alpha} \cdot p(s_j), \alpha = \epsilon, \lambda, 1$ $\overline{E}_i^{\alpha} = \sum_{j=1}^{n} \overline{u}_{ij}^{\alpha} \cdot p(s_j), \alpha = 1, \lambda, \epsilon$

For demonstrating the fact that fuzzy decision models offer the possibility to model real-world problems as realistic as a DM can describe them, the solution process is illustrated by using the following example:

Electro Fairies is a well known manufacturer of electric household utensils in Europe. Some years ago, *Electro Fairies* started to specialize in products with fuzzy control. The market rewards this new strategic alignment with the result that the current demand for these products is much higher than the production rate. Therefore, the management is confronted with the problem of making a decision on possible expansion of the production capacity. The management has the following possibilities:

a_1 : Renunciation of an enlargement of the capacity, status quo

a_2 : Enlargement of the actual manufacturing establishment by modernization of some plant with an increase in capacity of 30%

a_3 : Construction of a new plant with an increase in total capacity of 70%

a_4 : Construction of a new product line with an increase in total capacity of 100%

The profit earned by the different alternatives depends on the demand, which is not known with certainty. Due to the amount of information the manufacturer considers for the next 5 years either a "lower demand than actual" (state of nature s_1), "a constant demand" (state of nature s_2), "a higher demand" (state of nature s_3) or "a very high demand" (state of nature s_4). The manufacturer assigns the following prior probabilities to the states of nature:

$$p(s_1) = 0.30, \quad p(s_2) = 0.25, \quad p(s_3) = 0.35, \quad p(s_4) = 0.10.$$

The succeeding matrix of profits $\tilde{U}_{ij}$ displays which profits (measured in million Euro) correspond to the alternative constellations of output and demand. In order to avoid the problem of specifying utility values we assume risk neutrality. Under this assumption it does not influence the decision whether expected profits or expected utilities are applied and the Bernoulli criterion "maximization of the expected utility" is equivalent to the criterion "maximization of the expected profit."

Table 1. Prior profit matrix $\tilde{U}_{ij} = \tilde{U}(a_i, s_j)$

	s_1	s_2	s_3	s_4
a_1	(40; 42; 45; 45; 47; 50)	(50; 50; 50; 50; 50; 50)	(50; 50; 50; 50; 50; 50)	(50; 50; 50; 50; 50; 50)
a_2	(-5; 0; 5; 10; 15; 20)	(75; 77; 80; 85; 88; 90)	(80; 85; 90; 95; 100; 105)	(95; 100; 105; 110; 115; 120)
a_3	(-40; -30; -25; -20; -15; -10)	(30; 35; 40; 50; 55; 60)	(110; 115; 120; 130; 135; 140)	(140; 150; 160; 165; 170; 180)
a_4	(-90; -85; -80; -70; -65; -60)	(0; 5; 10; 20; 25; 30)	(75; 77; 80; 85; 88; 90)	(170; 180; 190; 210; 220; 230)

Using the formula (10) the fuzzy expected profits of the 4 alternatives are:

Table 2. Expected profit matrix $\tilde{E}(a_i)$

	expected profit
a_1	(47; 47.6; 48.5; 48.5; 50; 50)
a_2	(54.75; 59; 63.5; 68.5; 73; 77.25)
a_3	(48; 55; 60.5; 68.5; 73.5; 79)
a_4	(16.25; 20.7; 25.5; 34.75; 39.55; 44)

The question, which action a_i, $i = 1, 2, 3, 4$, is the best one, is not as trivial as in the classical model because of the fact that fuzzy sets are not well ordered. Comparing the membership functions of the expected profits in Figure 2, one can suppose that the alternatives

a_4 and a_1 turn out a lot worse than the alternatives a_2 and a_3. But for specifying an order of precedence it is necessary to define a preference relation between fuzzy sets.

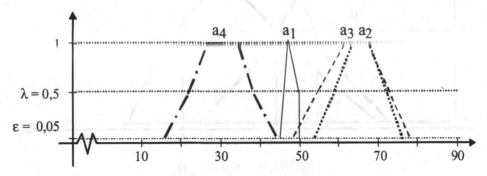

Figure 2. Membership functions of the expected profits

4 Fuzzy Preference Orderings

In the literature, various concepts for comparing fuzzy sets and for constructing preference orderings have been proposed (see e.g. Dubois, Prade 1983, Bortolan, Degani 1985, Rommelfanger 1986, Lai, Hwang 1992). Most of them are based on defuzzification, which means that each fuzzy set is compressed into a single crisp real number and the preference ordering is based on these crisp numbers, neglecting the spreads of the fuzzy sets. Therefore, only the following preference criteria, the ρ-preference and the ε-preference, are essential.

ρ-preference:
A fuzzy set $\tilde{B}$ is preferred to a fuzzy set $\tilde{C}$ on the ρ-level, $\rho \in [0,1]$, written as $\tilde{B} \succ_\rho \tilde{C}$, if

- ρ is the least real number, such that

$$\text{Inf } B_\alpha \geq \text{Sup } C_\alpha \quad \text{for all } \alpha \in [\rho, 1] \tag{11}$$

- and for at least one $\alpha \in [\rho, 1]$ the inequality (11) holds in the strict sense.

Here, $B_\alpha = \{x \in X \mid \mu_{\tilde{B}}(x) \geq \alpha\}$ and $C_\alpha = \{x \in X \mid \mu_{\tilde{C}}(x) \geq \alpha\}$ are the α-level-sets of $\tilde{B}$ and $\tilde{C}$, see Figure 3.

Since fuzzy intervals of the ε-λ-type are precisely specified only on the levels ε, λ and 1, it is wise to restrict the preference observations only to these three membership degrees.

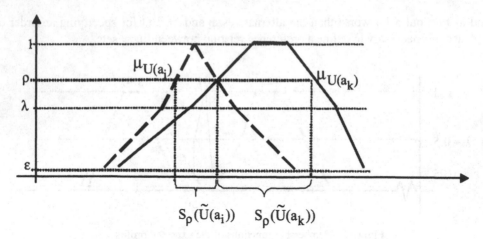

$$S_\rho(\tilde{U}(a_i)) \quad S_\rho(\tilde{U}(a_k))$$

Figure 3. Membership functions of the sets $\tilde{B}$ and $\tilde{C}$; ρ-preference

Looking at Figure 2 we observe the following ρ-preference relations:
If $\rho = \varepsilon$ holds, only the following relations are valid:

$a_2 \succ_\rho a_1$; $a_2 \succ_\rho a_4, a_3 \succ_\rho a_4, a_1 \succ_\rho a_4$, i.e. the alternative a_4 is worse than all other alternatives. Moreover $a_3 \succ_\rho a_1$ on a level about $\rho = 0,2$.

With regard to the ρ-preference it is not possible to decide between a_2 and a_3 .

Similar to this example, in many applications the ρ-preference relation does not lead to a preference ordering of the given alternatives. The cause of this disadvantage is the pessimistic attitude of the ρ-preference where the left part of the membership function of the better alternative is compared with the right-hand side of the membership function of the worse alternative. Therefore, the following preference relation is more appropriate and suitable for application in decision models.

ε-preference:
A fuzzy set $\tilde{B}$ is preferred to a fuzzy set $\tilde{C}$ on the level $\varepsilon \in [0,1]$, written as $\tilde{B} \succ_\varepsilon \tilde{C}$, if

- ε is the least real number, such that
  $$\text{Sup } B_\alpha \geq \text{Sup } C_\alpha \text{ and Inf } B_\alpha \geq \text{Inf } C_\alpha \text{ for all } \alpha \in [\varepsilon, 1] \tag{12}$$

- and for at least one $\alpha \in [\varepsilon, 1]$ one of these inequalities holds in the strict sense.

For fuzzy intervals $\tilde{X}_i = (\underline{x}_i^\varepsilon, \underline{x}_i^\lambda, \underline{x}_i^1, \bar{x}_i^1, \bar{x}_i^\lambda, \bar{x}_i^\varepsilon)^{\varepsilon,\lambda}$ of the ε-λ-type the terms (12) can be simplified to

$$\tilde{X}_i \succ_\varepsilon \tilde{X}_j \iff \underline{x}_i^\alpha > \underline{x}_j^\alpha \text{ and } \bar{x}_i^\alpha > \bar{x}_j^\alpha \quad \text{for } \alpha = \varepsilon, \lambda, 1 \tag{13}$$

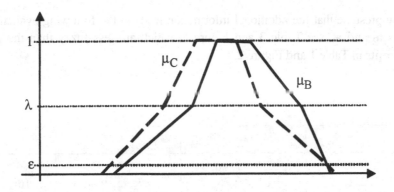

Figure 4. Membership functions of the sets $\widetilde{B}$ and $\widetilde{C}$; ε-preference

As per definition $\operatorname{Sup} B_{\alpha} \geq \operatorname{Inf} B_{\alpha}$ and $\operatorname{Sup} C_{\alpha} \geq \operatorname{Inf} C_{\alpha}$ the terms (12) are direct deductions of the term (11), i.e. the ε-preference relation is weaker than the ρ-preference. Therefore, the ε-preference should only be used on the level ε. Using the ε-preference the following preference orderings are valid: $a_2 \succ_{\varepsilon} a_1 \succ_{\varepsilon} a_4$, $a_3 \succ_{\varepsilon} a_1 \succ_{\varepsilon} a_4$. But even with the ε-preference it is not possible to decide between a_2 and a_3.

5 The Use of Additional Information

In real-world applications the DM is often not able to specify objective prior probability distributions $p(s_j)$, $j = 1,...,n$. For improving the situation the DM could look for additional information $X = \{x_1, x_2,...,x_K\}$. Knowing the likelihoods $p(x_k \mid s_j)$, the prior probability distribution $p(s_j)$ can be substituted by the posteriori probability distribution: see Section 2.

However the calculation of posterior probabilities is a complicated procedure which needs a lot of information and implies intensive calculations. In practice the DM has to devote money and time to these activities, before it is actually possible to calculate the value of additional information X. In particular, it seems absolutely improbable that a DM knows all likelihoods. Empirical opinion polls indicate that posterior probabilities are not applied in case of solving real decision problems.

Instead of using posterior probabilities in fuzzy systems the DM can as well try to specify the results associated with act–event combinations more precisely by gaining additional information.

In Table 2 and Figure 4 it becomes evident that, assuming the a priori distribution is correct, the alternative a_4 as well as the alternative a_1 will not be taken into consideration furthermore. Therefore, additional information should be gathered only about the alternatives a_2, a_3 for getting profit values which are less fuzzy, i.e. have smaller 1-level-sets or smaller spreads.

We now presume that the additional information leads to the following evaluations of the alternatives a_2 and a_3, see Table 3 and Figure 5, which are less fuzzy than the profits and expected profits in Table 1 and Figure 2.

Table 3. Prior profit matrix with additional information $\tilde{U}^I(a_i, s_j)$

	s_1	s_2	s_3	s_4
a_2	(2; 5; 8; 10; 12; 15)	(75; 77; 80; 83; 85; 88)	(87; 90; 93; 93; 95; 100)	(102; 105; 107; 109; 112; 115)
a_3	(-38; -29; -25; -22; -20; -15)	(38; 40; 44; 46; 50; 53)	(115; 118; 121; 123; 125; 130)	(150; 156; 161; 163; 165; 170)

Table 4. Expected profit matrix with additional information $\tilde{E}_U^I(a_i)$

	expected profit
a_2	(60; 62.75; 65.65; 67.2; 69.3; 73)
a_3	(53.35; 58.2; 61.95; 64.25; 66.75; 71.25)

Regarding the ε-preference, Figure 5 indicates that $a_2 \succ_\varepsilon a_3 \succ_\varepsilon a_1 \succ_\varepsilon a_4$, i.e. the alternative a_2 is the best one. This identification is confirmed by $a_2 \succ_\rho a_1 \succ_\rho a_4$ and $a_3 \succ_\rho a_1 \succ_\rho a_4$ on the level $\rho = \varepsilon$.

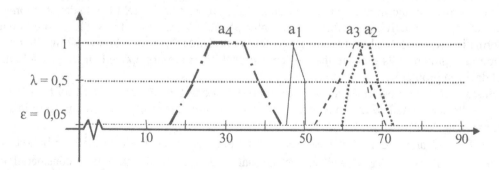

Figure 5. Membership functions of the expected profits with additional information

With the additional information the DM can be almost sure that the alternative a_2 is the best in accordance with the preference criterion "maximization of the expected utility." Nevertheless it is almost impossible to define the value of this information.

This course of calculation in which the values associated with act–event combinations are modeled by fuzzy intervals of ε-λ-type should be repeated, and thus the evaluations can be improved by gradually introducing additional information. The essential advantage of this interactive procedure is that it presents an adequate answer to the information dilemma of real problems.

In order to limit the extensive information process one could start to design a model of the real problem applying only such information, which can be obtained with little effort and at reasonable cost. In case of modeling by fuzzy intervals one has to accept the disadvantage that some of the parameters show large spreads. In general we will get no clear ranking of the alternatives, but usually we can observe that only few actions are taken into consideration. Only the evaluations of these decisive alternatives should be improved by collecting additional information. By doing so, the costs for additional information can be reduced. Unlike the extensive gathering of information ex ante -which is inevitable in classical models - the acquisition of additional information will then be designed in accordance with the set aims and carried out under consideration of cost-benefit relations.

6 Fuzzy Probabilities

It has to be considered that detailed information about the occurrence of the states of nature may not be available. As a consequence it could occur that the prior probabilities are not described precisely, but only vaguely by means of fuzzy intervals of the ε-λ-type

$$\tilde{P}(s_j) = (\underline{p}_j^\varepsilon; \underline{p}_j^\lambda; \underline{p}; \overline{p}_j^l; \overline{p}_j^\lambda; \overline{p}_j^\varepsilon)^{\varepsilon,\lambda} , j = 1, 2,..., n.$$

For our example we assume that the fuzzy probabilities in Table 5 are given:

Table 5. Fuzzy probabilities $\tilde{P}(s_j)$

s_j	$\tilde{P}(s_j) = (\underline{p}_j^\varepsilon; \underline{p}_j^\lambda; \underline{p}_j^l; \overline{p}_j^l; \overline{p}_j^\lambda; \overline{p}_j^\varepsilon)^{\varepsilon,\lambda}$
s_1	$\tilde{P}(s_1) = (0.26; 0.28; 0.29; 0.3; 0.31; 0.33)^{\varepsilon,\lambda}$
s_2	$\tilde{P}(s_2) = (0.20; 0.23; 0.25; 0.25; 0.27; 0.30)^{\varepsilon,\lambda}$
s_3	$\tilde{P}(s_3) = (0.30; 0.32; 0.33; 0.36; 0.38; 0.41)^{\varepsilon,\lambda}$
s_4	$\tilde{P}(s_4) = (0.08; 0.09; 0.1; 0.1; 0.11; 0.13)^{\varepsilon,\lambda}$

The following term offers a simple approximation method for calculating the fuzzy expected profits:

$$\tilde{E}_i^A = (\underline{E}_i^\varepsilon; \underline{E}_i^\lambda; \underline{E}_i^1; \overline{E}_i^1; \overline{E}_i^\lambda; \overline{E}_i^\varepsilon)^{\varepsilon,\lambda} = \tilde{U}_{i1} \cdot \tilde{P}(s_1) \oplus \cdots \oplus \tilde{U}_{in} \cdot \tilde{P}(s_n) \tag{14}$$

where $\underline{E}_i^\alpha = \sum\limits_{J=1}^n \underline{u}_{ij}^\alpha \cdot \underline{p}_j^\alpha$, $\alpha = \varepsilon, \lambda, 1$ $\qquad \overline{E}_i^\alpha = \sum\limits_{J=1}^n \overline{u}_{ij}^\alpha \cdot \overline{p}_j^\alpha$, $\alpha = 1, \lambda, \varepsilon$

At first we discuss the special case that the profits are crisp. Moreover we assume that the DM is risk neutral and the profits are given by Table 6.

Table 6. Profits (measured in million Euro)

	s_1	s_2	s_3	s_4
a_1	50	50	50	50
a_2	10	82	93	108
a_3	-22	45	123	155
a_4	-75	15	82	200

The application of formula (14) on the crisp profits in Table 6 and the fuzzy probabilities in Table 5 lead to the fuzzy expected profits in Table 7:

Table 7. Fuzzy expected profits $\tilde{E}_i^A$

	$\tilde{E}_i^A = (\underline{E}_i^\varepsilon; \underline{E}_i^\lambda; \underline{E}_i^1; \overline{E}_i^1; \overline{E}_i^\lambda; \overline{E}_i^\varepsilon)^{\varepsilon,\lambda}$
a_1	(42; 46; 48.5; 50.5; 53.5; 58.5)
a_2	(55.54; 61.14; 64.89; 67.78; 72.46; 80.07)
a_3	(52.58; 57.5; 60.96; 64.43; 69.12; 76.82)
a_4	(24.10; 26.69; 29.06; 30.77; 33.96; 39.37)

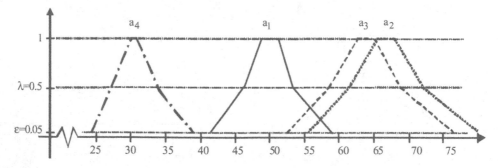

Figure 6. Membership functions of fuzzy expected profits $\tilde{E}_i^A$

Figure 6 reveals that according to the $\succ_\varepsilon$-preference the alternative a_2 is best within the set considered.

According to the $\succ_\rho$-preference the alternatives a_2 and a_3 are better than the alternative a_4 on the level $\rho = \varepsilon$ and better than the alternative a_1 on a level $\beta < 0.5$.

Using the formula (14) for calculating approximately the fuzzy expected profits the important restriction $\sum_{j=1}^{n} p_j = 1$ was neglected. During the calculation of the precise expected values $\tilde{E}_i^P$ this constraint is observed and the following statement is true:

$$\tilde{E}_i^P \subseteq \tilde{E}_i^A \Leftrightarrow \mu_{E_i^P}(u) \leq \mu_{E_i^A}(u). \tag{15}$$

It follows that the $\succ_\rho$-preference orders on the level $\rho = \varepsilon$ remain valid if the values $\tilde{E}_i^A$ are replaced by $\tilde{E}_i^P$. Therefore, the calculation of $\tilde{E}_i^P$ is only necessary for those alternatives that are not definitively excluded on the basis of $\tilde{E}_i^A$'s.

For calculating the fuzzy expected profits $\tilde{E}_i^P = (\underline{E}_i^\varepsilon; \underline{E}_i^\lambda; \underline{E}_i^1; \overline{E}_i^1; \overline{E}_i^\lambda; \overline{E}_i^\varepsilon)^{\varepsilon,\lambda}$ we can use the following terms:

$$\underline{E}_i^\alpha = \text{Min}\{ \sum_{j=1}^{n} \underline{u}_{ij}^\alpha \cdot p_j | p_j \in [\underline{p}_j^\alpha, \overline{p}_j^\alpha] \text{ and } \sum_{j=1}^{n} p_j = 1\} \quad \alpha = \varepsilon, \lambda, 1 \tag{16}$$

$$\overline{E}_i^\alpha = \text{Max}\{ \sum_{j=1}^{n} \overline{u}_{ij}^\alpha \cdot p_j | p_j \in [\underline{p}_j^\alpha, \overline{p}_j^\alpha] \text{ and } \sum_{j=1}^{n} p_j = 1\} \quad \alpha = 1, \lambda, \varepsilon \tag{17}$$

For simplifying the calculation of $\tilde{E}_i^P$, the following algorithms can be used:

Algorithm for the calculation of auxiliary probabilities $\underline{p}_j^\alpha(i)$ for calculating $\underline{E}_i^\alpha$, $\alpha = \varepsilon, \lambda, 1$.

1. Specify for all probabilities the smallest value: $\underline{p}_j^\alpha(i) = \underline{p}_j^\alpha$.

2. Increase the probabilities of the state of nature with the smallest utility value. If this is given for s_n, we will get

$$\underline{p}_n^\alpha(i) = \text{Max}\{p \in [\underline{p}_n^\alpha, \overline{p}_n^\alpha] | \sum_{j=1}^{n-1} \underline{p}_j^\alpha + p \leq 1\} \tag{18}$$

3. If the inequality is fulfilled in the strong sense, we will increase the probability of the state of nature with the second smallest utility value. If this is given for s_{n-1}, we will make the calculation

$$\underline{p}_{n-1}^\alpha(i) = \text{Max}\{p \in [\underline{p}_{n-1}^\alpha, \overline{p}_{n-1}^\alpha] | \sum_{j=1}^{n-2} \underline{p}_j^\alpha + p + p_n^\alpha \leq 1\} \tag{19}$$

4. This procedure is shall be continued if the inequality (18) is not fulfilled as equation, and so on.

Algorithm for the calculation of auxiliary probabilities $\overset{=\alpha}{p}_j$ (i) for calculating $\overline{E}_i^\alpha$, $\alpha = 1, \lambda, \varepsilon$.

1. Specify for all probabilities the smallest value: $\overset{=\alpha}{p}_j$ (i) $= \underset{=j}{p^\alpha}$.

2. Increase the probabilities of the state of nature with the highest utility value. If this is given for s_1 , we will get

$$\overset{=\alpha}{p}_1 \text{ (i)} = \text{Max}\{p \in [\underline{p}_1^\alpha, \overline{p}_1^\alpha] \mid p + \sum_{j=2}^{n} \underset{-j}{p^\alpha} \leq 1\} \tag{20}$$

3. If the inequality is fulfilled in the strong sense, we will increase the probability of the state of nature with the second highest utility value. If this is given for s_2, we will make the calculation

$$\overset{=\alpha}{p}_2 \text{ (i)} = \text{Max}\{p \in [\underline{p}_2^\alpha, \overline{p}_2^\alpha] \mid p_1^\alpha + p + \sum_{j=3}^{n} \underset{-j}{p^\alpha} \leq 1\} \tag{21}$$

4. This procedure shall be continued if the inequality (20) is not fulfilled as equation, and so on..

If we apply these algorithms to the example with crisp profit values we receive the probabilities $\overset{=\alpha}{p}_j$ (i) and $\underset{=j}{p^\alpha}$(i) , which are in this special example independent of i, because the profits in Table 6 comply with the ordering relation $x_{i1} \leq x_{i2} \leq x_{i3} \leq x_{i4}$ for all i = 1, 2,3, 4.

Table 8. Auxiliary probabilities $\overset{=\alpha}{p}_j$ (i) and $\underset{=j}{p^\alpha}$(i)

	ε	λ	1		1	λ	ε
$\underset{=1}{p^\alpha}$	0.33	0.31	0.30	$\overset{=\alpha}{p}_1$	0.29	0.28	0.26
$\underset{=2}{p^\alpha}$	0.29	0.27	0.25	$\overset{=\alpha}{p}_2$	0.25	0.23	0.20
$\underset{=3}{p^\alpha}$	0.30	0.33	0.35	$\underset{=3}{p^\alpha}$	0.36	0.38	0.41
$\overset{=\alpha}{p}_4$	0.08	0.09	0.10	$\overset{=\alpha}{p}_4$	0.10	0.11	0.13

With these auxiliary probabilities $\overset{=\alpha}{p}_j$ and $\underset{=j}{p^\alpha}$ the fuzzy expected profits $\tilde{E}_i^P$ can be calculated as in Table 9:

Table 9. Fuzzy expected profits $\tilde{E}_i^P$

	$\tilde{E}_i^P = (\underline{E}_i^\varepsilon; \underline{E}_i^\lambda; \underline{E}_i^1; \overline{E}_i^1; \overline{E}_i^\lambda; \overline{E}_i^\varepsilon)^{\varepsilon, \lambda}$
a_1	$(50,50, 50, 50, 50, 50)$
a_2	$(63.62; 65.65; 66.85; 67.68; 68.88; 71.17)$
a_3	$(55.09; 59.87; 63.2; 64.65; 67.98; 73.86)$
a_4	$(20.20; 25.86; 29.95; 31.52; 35.61; 43.12)$

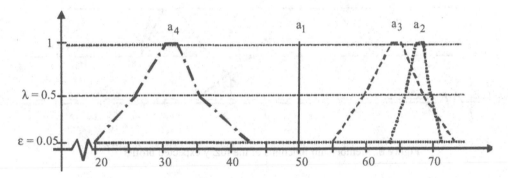

Figure 7. Membership functions of the fuzzy expected profits $\tilde{E}_i^P$

Comparing the shapes of the membership functions of the $\tilde{E}_i^P$ in Figure 7 with those of $\tilde{E}_i^A$ in Figure 8 we can clearly recognize that the fuzzy values $\tilde{E}_i^P$ are less fuzzy than their approximations $\tilde{E}_i^A$. As a result the ranking order $a_2 \succ_\rho a_3 \succ_\rho a_1$ is now secure on a level $\rho \approx 0,75$ between a_2 and a_3 and the level $\rho = \varepsilon$ between a_3 and a_1. Moreover the expected profit $\tilde{E}_4^P$ is now crisp (as requested), because all act-event-profiles of a_4 have the same value 50.

Finally we will calculate fuzzy expected profits $\tilde{E}_i^P$ for the general case that not only the probabilities are fuzzy but also the profits. For that the terms (16) and (15) and the algorithms for calculating auxiliary probabilities can be used as they are developed for the general case. Using the fuzzy profits of Table 1 (for a_1 and a_4), Table 3 (for a_2 and a_3) and the auxiliary fuzzy probabilities of Table 8, the expected profits $\tilde{E}_i^P$ as in Table 10 are calculated.

Table 10. Fuzzy expected profits $\tilde{E}_i^P$

$\tilde{E}_i^P = (\underline{E}_i^\varepsilon; \underline{E}_i^\lambda; \underline{E}_i^1; \overline{E}_i^1; \overline{E}_i^\lambda; \overline{E}_i^\varepsilon)^{\varepsilon,\lambda}$
a_1	(46.7; 47.52; 48.5; 48.55; 50; 50)
a_2	(56.67; 61.49; 65.65; 68.03; 71.33; 77.45)
a_3	(44.98; 54.79; 61.95; 65.70; 71.55; 82.10)
a_4	(6.40; 16.61; 25.5; 36.3; 45.19; 57.20)

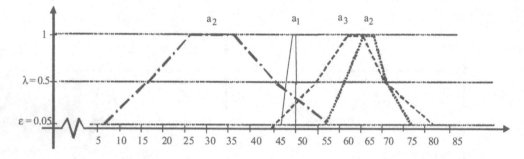

Figure 8. Membership functions of the fuzzy expected profits $\tilde{E}_i^P$

Figure 8 reveals that based on the ε-preference even this imprecise information is sufficient to exclude alternative a_4 and a_1 from further consideration, whereas neither the ε-preference nor the ρ-preference is adequate for establishing a preference order between a_2 and a_3. In our opinion the preference order $a_2 \succ_\rho a_3$ on the level $\varepsilon = \lambda$ is not sufficient. In order to come to a decision additional information should be collected to get a more precise description of the values of the remaining alternatives or the probabilities $\tilde{P}(s_j)$.

7 Conclusions

By applying fuzzy intervals, real decision problems can be modeled as exactly as the DM wants to or can perform. In doing so, the DM does not run the risk of choosing an alternative that provides an optimal solution for the model, but does not match the real problem.

A disadvantageous consequence of the use of fuzzy results or fuzzy probabilities is the fact that a best alternative is not identified in all applications. But normally it is possible to reject the majority of the alternatives because they have a worse ranking compared with the remaining alternatives concerning the ε- or the ρ-preference. In order to reach a ranking of the remaining alternatives additional information about the results of these alternatives can be used.

Apart from the fact that fuzzy models offer a more realistic modeling of decision situations, the proposed solution process leads to a reduction of information costs. This circumstance is caused by the fact that additional information is gathered in correspondence with the requirements and under consideration of cost-benefit-relations. It is recommended to refrain from collecting expensive additional information a priori and to start with the information that the DM has or can get with low costs.

References

Bortolan, G., and Degani, R. (1985). Ranking fuzzy subsets. *Fuzzy Sets and Systems* 15: 1-19.

Chen S.J. and Hwang Ch.-L. (1987). *Fuzzy Multiple Attribute Decision Making, Methods and Applications*. Berlin: Heidelberg: Springer Verlag.

Dubois, D., and Prade, H. (1982). The Use of Fuzzy Numbers in Decision Analysis. In (Gupta, M.M. and Sanchez, E. Hrsg.) *Fuzzy Information and Decision Processes*. Amsterdam New York Oxford, 309–321.

Dubois, D,. and Prade, H. (1983). Ranking of Fuzzy Numbers in the Setting of Possibility Theory. *Information Sciences* 30: 183-224.

Dubois, D,. and Prade, H. (1998).Possibilistic logic in decision, In *Proceedings of EFDAN'98*. 41-49

Fandel, G., Francois, P., and Gulatz, K.M: (1994). *PPS-Systeme: Grundlagen, Methoden, Software, Marktanalyse*. Heidelberg

Jain, R. (1976). Decision making in the presence of variables. *IEEE, Transactions on Systems, Man and Cybernetics* 6: 698-703.

Kivijärvi, H., Korhonen, P., and Wallenius, J. (1986). Operations research and its practice in Finland. *Interfaces* 16: 53-59.

Lai, Y.-J., and Hwang, C.-L. (1992). *Fuzzy mathematical programming*. Berlin Heidelberg: Springer.

Lilien, G. (1987). MS/OR: A mid-life crises. *Interfaces* 17: 53-59

Menges, G., and Kofler, E. (1976). *Entscheidungen bei unvollständiger Information*. Berlin Heidelberg: Springer.

Meyer zu Selhausen H. (1989). Repositioning OR's Products in the Market. *Interfaces* 19, 79–87

Neumann, J.v., and Morgenstern, O. (1953). *Theory of games and economic behavior*. Princeton.

Ramik, J. und Rimanek, J. (1985). Inequality between Fuzzy Numbers and its Use in Fuzzy Optimization. *Fuzzy Sets and Systems* 16: 123-138.

Rommelfanger, H. (1984). Entscheidungsmodelle mit Fuzzy-Nutzen. In *Operations Research Proceedings* 1983, 559-567

Rommelfanger H. (1986). Rangordnungsverfahren für unscharfe Mengen. *OR-Spektrum* 8: 219–228

Rommelfanger H. (1994). *Fuzzy Decision Support-Systeme – Entscheiden bei Unschärfe*. Berlin, Heidelberg: Springer, 2nd edition.

Rommelfanger, H., and Eickemeier, S. (2002). *Entscheidungstheorie. Klassische Konzepte und Fuzzy-Erweiterungen*. Berlin Heidelberg: Springer.

Slowinski, R. (Ed.) (1998). *Fuzzy Sets in Decision Analysis, Operations Research and Statistics*. Boston: Kluwer.

Sommer, G. (1980). *Bayes-Entscheidungen mit unscharfer Problembeschreibung*. Frankfurt am Main: Peter Lang

Tanaka H., Okuda T., and Asai K. (1976). A Formulation of Fuzzy Decision Problems and its Application to an Investment Problem. *Kybernetes* 5: 25–30.

Tanaka, H., and Asai, K. (1984). Fuzzy Linear Programming with Fuzzy Numbers. *Fuzzy Sets and Systems* 13: 1-10.

Tanaka, H., Ichihashi, H., and Asai, K. (1984). A Formulation of Linear Programming Problems based on Comparison of Fuzzy Numbers. *Control and Cybernetics* 13: 185-194.

Tingley, G.A. (1987). Can MS/OR sell itself well enough? *Interfaces* 17: 41-52.

Watson, S.R., Weiss, J.J., and Donell, M.L. (1979). Fuzzy Decision Analysis. *IEEE, Transactions on Systems, Man and Cybernetics* 9: 1-9.

Whalen, T. (1984). Decision Making under Uncertainty with various Assumptions about available Information. *IEEE, Transactions on Systems, Man and Cybernetics* 14: 888-900.

Yager, R.R. (1979). Possibilistic Decision Making. *IEEE, Transactions on Systems, Man and Cybernetics* 9:

Zadeh L.A. (1965). Fuzzy Sets. *Information and Control* 8: 338–353

Zimmermann H.J. (1996). *Fuzzy Sets Theory and its Applications*. Boston: Kluwer.

Capital Allocation under Regret and Kataoka Criteria

Günter Bamberg [‡] and Gregor Dorfleitner [‡]

[‡] Wirtschaftswissenschaftliche Fakultät der Universität Augsburg, Augsburg, Germany

Abstract. The paper analyzes the allocation of a given initial capital between a risk-free and a risky alternative. Typically, the risky alternative is the investment into the stock market or into a stock market index. Under expected utility the optimal fraction a_* to invest into the stock market depends on the initial capital, on the distribution of stock returns, on the planning horizon, and of course on the von Neumann/Morgenstern utility function. Moreover, the optimal a_* can only be evaluated by numerical integration. In order to get explicit formulas and to avoid the problematic assessment of the utility function NEU (non expected utility) approaches are discussed. The maxmin and the minmax regret criterion select only corner solutions (i.e. $a_* = 0$ or $a_* = 1$). The following Kataoka variant of these criteria is considered: Fix a (small) probability α and discard all the extremal events (which have althogether the probability α) from the planning procedure; i.e. define the worst case by exclusion of these extremal events. Obviously, this idea is also the basis of the well-known value-at-risk approach. The optimal fraction a_* is no longer a corner solution. Moreover, it allows explicit formulas. These are studied in the Black/Scholes world (i.e. normally distributed log returns). Under realistic parameter values a_* increases with the length of the planning horizon.

1 Introduction

Two-fund separation suggests the analysis of portfolios which are structured as follows: A fraction a of the initial capital $v(0)$ is to be invested into the stock market whereas the rest, i.e. the amount $(1 - a) \cdot v(0)$, is to be invested at the risk-free rate r. Final wealth at planning horizon T stemming from such portfolios is

$$V(T) = v(0) \left[ae^{R(T)} + (1 - a)e^{rT} \right], \tag{1.1}$$

where both the (annualized) risk-free rate r and the (non-annualized) stochastic rate $R(T)$ are compounded continuously. These portfolios are not only suggested by two-fund separation but offer the following advantages:

- They are very easy to implement. Investment into the stock market could be realized by purchasing index certificates or index funds, in Germany for instance DAX-Participations.
- They provide broad diversification and avoid the taking of unsystematic risks.

- No reallocation from now $(t = 0)$ until the planning horizon $(t = T)$ is required. Therefore, the investor does not incur additional transaction costs.

Admittedly, it is nowadays (end of 2002) very unpopular to consider an investment into the stock market. On the other hand, empirical data provide strong evidence that the mean return is higher on the stock market than on a bank account or on treasury bills. And it is a trivial truism that the return on the stock market is even higher if the investor succeeds to buy at or near the low.

We assume that the initial capital is an amount which is at the investor's disposal. The amount is superfluous in the sense that the investor can afford to sit out the whims of the market. Hence, liquidity constraints are irrelevant. For the same reason we restrict the fraction a to the interval $[0; 1]$, thus excluding short sales $(a < 0)$ on the stock market and purchase of stocks on credit $(a > 1)$.

Only a, the initial fraction invested into shares, is subject to optimization. We address the questions:

- What is the optimal fraction a_*?
- How sensitive does the optimal fraction a_* react on the choice of the decision rule?
- How does a_* depend on the planning horizon T?

Bamberg, Dorfleitner and Lasch (1999) analyzed these questions in the expected utility framework. For HARA utility functions they found that a_* is an increasing function of the planning horizon T. Even in this restricted class of utility functions numerical integration was required to calculate the optimal fraction. Undoubtedly, expected utility is a valuable workhorse for theoretical purposes. But practitioners, especially in the field of financial consulting, are putting forward serious objections against the use of expected utility. They point to the problematical assessment of the utility function and to the difficult communication with non-academic clients. It is therefore tempting to use "quick and dirty" decision procedures which do not require more information the investor is willing or able to provide. There are several papers investigating the optimal fraction a_* in the framework of shortfall models, e.g. Bamberg, Dorfleitner and Lasch (1999), Wolter (1993), Zimmermann (1991). In the sequel we study the optimal fraction under maxmin and minmax regret criteria and Kataoka variants of these criteria.

There is a variety of models dealing with the fraction(s) to invest riskily. At first glance they look rather similar. However, they are very different and it is not possible to draw inference from one model to another. To get an idea of the model variety, let the time be discrete, thus actions can be taken only at $t = 0, 1, \ldots, T - 1$.

- In the **model under consideration** merely one decision variable matters, namely $a(= a_0)$. The decision maker is a buy and hold investor. The fraction a_t riskily invested at time t,

$$a_t = \frac{a \cdot e^{R(t)}}{a \cdot e^{R(t)} + (1 - a)e^{rt}} \ ,$$

is random; in particular it is neither constant nor investor controlled.

- In **fixed-proportions-through-time** models there is only one decision variable as well, namely $a(= a_0 = a_1 = \ldots = a_{T-1})$. The riskily invested fraction is kept constant, thus requiring permanent reallocations and high transaction costs.
- In **market timer** models there are T decision variables $a_0, a_1, \ldots, a_{T-1}$, one for each period. Conclusions must be drawn dependent on the return history (which is problematic in a rather efficient capital market).
- Finally, in **consumption oriented** models there are $2T$ decision variables, namely the T fractions $a_0, a_1, \ldots, a_{T-1}$ and the T period consumptions $c_0, c_1, \ldots, c_{T-1}$. In models of this type the objective function does not only refer to final wealth but also to the level and smoothness of the consumption pattern.

2 Distribution of log return $R(T)$

According to the Black/Scholes world $R(T)$ is normally distributed with

$$E[R(T)] = \left(\mu - \frac{\sigma^2}{2} \right) T \quad , \quad \text{Var}\,[R(T)] = \sigma^2 T \tag{2.1}$$

(μ = drift, σ = volatility). Hence stock price (index)

$$S(T) = S(0)e^{R(T)}$$

is lognormal with expected value

$$E[S(T)] = S(0)e^{\mu T} \tag{2.2}$$

and final wealth $V(T)$ is lognormal (up to a righthand shift of $v(0)(1-a)e^{rT}$). Empirical data suggest (long-term, annualized) parameter values

$$\sigma \approx 0.20 \quad ; \quad \mu \approx 0.10 \quad ; \quad \mu - \frac{\sigma^2}{2} - r \approx 0.04 . \tag{2.3}$$

In particular, the risk premium (difference between expected log return and risk-free return) can be assumed positive:

$$\mu - \frac{\sigma^2}{2} - r > 0 .$$

For instance, US data for the 71 years from 1926 to 1996 (taken from Lo and MacKinlay (1998)) are

$$\text{stocks:} \quad \frac{S(1996)}{S(1926)} = 1370 \quad ; \quad \text{treasury bills:} \quad \frac{S(1996)}{S(1926)} = 14.$$

These data and (2.2) yield the crude estimates

$$\hat{\mu} = \frac{1}{71} \ln 1370 = 10.17\% \quad ; \quad \hat{r} = \frac{1}{71} \ln 14 = 3.72\% \, .$$

Since the paper relies on normally distributed (and thus thin-tailed) log returns the question of alternative distributions is a very legitimate issue. Many papers and monographs, for instance Krämer and Runde (1996) or Rachev and Mittnik (2002) and the references therein, are discussing stable (non-normal) distributions or other fat-tailed distributions like $t-$distributions or unconditional distributions stemming from ARCH or GARCH processes. Unfortunately, the fat-tailed distributions prevent the existence of the exponential moment $E[e^{R(T)}]$. Therefore neither expected prices

$$E[S(T)] = S(0)E\left[e^{R(T)}\right]$$

nor expected (simple) returns

$$E\left[\frac{S(T) - S(0)}{S(0)}\right] = E\left[e^{R(T)}\right] - 1$$

can be finite. As pointed out by Bamberg and Dorfleitner (2002) fat-tailed log returns destroy the underpinnings of the Markowitz world, CAPM etc. In the same paper it is shown that normality of $R(T)$ follows from four plausible axioms. Therefore, we continue to hold on the normality assumption with respect to log returns $R(T)$.

3 The optimal fraction a_* heavily depends on the decision criterion

Let $a_*(T)$ denote the optimal fraction as a function of the planning horizon T:

- Under **expected utility** one can show (compare Bamberg, Dorfleitner and Lasch (1999)): For investors with finite risk aversion it can never be optimal to invest all the money at the risk-free rate r, i.e.

$$a_*(T) > 0 \quad \text{for all} \quad T.$$

  For investors with constant relative risk aversion ρ (HARA class) and plausible numerical values with respect to μ, σ, r, ρ the optimal fraction $a_*(T)$ increases monotonically with the planning horizon T.
- **Shortfall models** have an inherent tendency towards higher fractions a. If, for instance, the required benchmark return exceeds the risk-free rate r, then $a = 0$ (i.e. only risk-free investment) entails a shortfall probability of 100%. Thus the investor is forced to engage excessively on the stock market.
- According to the **maxmin criterion** one has to look for the min (or inf) of (1.1). Since the support of (the normally distributed) $R(T)$ is $\mathbb{R}$, the worst case is $R(T) \to -\infty$ and the resulting final wealth is

$$v(0)(1 - a)e^{rT}. \tag{3.1}$$

Obviously, $a = 0$ maximizes (3.1). Therefore, the maxmin criterion gives the advice to invest all the money risk-free (and thus giving away any chance to earn an excess profit):

$$a_*(T) = 0 \quad \text{for all} \quad T. \tag{3.2}$$

- The (minmax) **regret criterion** gives just the opposite advice,

$$a_*(T) = 1 \quad \text{for all} \quad T. \tag{3.3}$$

This can be seen as follows: The regret related to decision a (contingent on $R(T)$) is

$$\max_{\tilde{a}} \quad V[\tilde{a}, R(T)] - V[a, R(T)] =$$

$$= \begin{cases} v(0)e^{R(T)} - v(0)\left[ae^{R(T)} + (1-a)e^{rT}\right] & , \quad \text{if} \quad R(T) > rT \\ v(0)e^{rT} - v(0)\left[ae^{R(T)} + (1-a)e^{rT}\right] & , \quad \text{if} \quad R(T) \le rT \end{cases}$$

$$= \begin{cases} v(0)(1-a)\left[e^{R(T)} - e^{rT}\right] & , \quad \text{if} \quad R(T) > rT \\ v(0)a\left[e^{rT} - e^{R(T)}\right] & , \quad \text{if} \quad R(T) \le rT. \end{cases} \tag{3.4}$$

Obviously,

$$\max \text{regret}_{-\infty < R(T) < \infty} = \begin{cases} \infty & , \quad \text{if} \quad 0 \le a < 1 \\ v(0)e^{rT}(<\infty) & , \quad \text{if} \quad a = 1, \end{cases}$$

such that minmax regret is attained for $a = 1$. The worst case is now $R(T) \to +\infty$ bringing about the maximal regret.

Note that both (3.2) and (3.3) are corner solutions and that these solutions are determined through the most extreme (but very unlikely) stock prices. The next section investigates the consequences of cutting out astronomically high and catastrophically low stock prices.

4 Kataoka combined with maxmin and regret criteria

Kataoka (1963) proposed to fix a small probability α and to select the action which maximizes the α-quantile of the payoff distribution. Though this criterion is not compatible with expected utility, it has gained importance, maybe due to the fact that it is closely related to the well-known value-at-risk approach. The following (and slightly more general) routine is also termed after Kataoka:

Specify a (small) probability α and discard all the unfavorable extremal events (which have altogether the probability α) from the planning procedure.

Obviously, the original Kataoka criterion is a combination of the general Kataoka procedure with the maxmin criterion. To combine the general Kataoka procedure with the regret criterion the meaning of "unfavorable extremal events" has to be made precise. Since both extremely high and extremely low returns can bring about regret, we will discuss the truncation on one side and on both sides.

Kataoka combined with maxmin

The worst case is unambiguous: returns $R(T)$ or final wealth $V(T)$ are too low. Thus only returns greater or equal the α−quantile q are relevant,

$$R(T) \geq q = \left(\mu - \frac{\sigma^2}{2} \right) T + z(\alpha)\sigma\sqrt{T}, \tag{4.1}$$

where $z(\alpha)$ is the α−quantile of $N(0,1)$. The minimal final value,

$$\min \ \{V[a, R(T)] : R(T) \geq q\},$$

is

$$\begin{cases} v(0)\{e^{rT} + a[e^q - e^{rT}]\} & , \ \text{if} \ \ a > 0 \\ v(0)e^{rT} & , \ \text{if} \ \ a = 0 \end{cases} . \tag{4.2}$$

The fraction a which maximizes (4.2) depends on the sign of the difference

$$e^q - e^{rT}. \tag{4.3}$$

If (4.3) is positive (negative) then $a = 1$ ($a = 0$) is optimal. If (4.3) is equal to zero then a is arbitrary between 0 and 1. Setting (4.3) equal to zero is equivalent to

$$\left(\mu - \frac{\sigma^2}{2} \right) T + z(\alpha)\sigma\sqrt{T} = rT.$$

Solving with respect to the planning horizon T gives

$$\hat{T}(\alpha) = \left(\frac{z(\alpha)\sigma}{\mu - \frac{\sigma^2}{2} - r} \right)^2. \tag{4.4}$$

Thus the combination of Kataoka with maxmin yields the advice

$$a_*(T) = \begin{cases} 0 & , \ \text{if} \ \ T < \hat{T}(\alpha) \\ \text{arbitrary} & , \ \text{if} \ \ T = \hat{T}(\alpha) \\ 1 & , \ \text{if} \ \ T > \hat{T}(\alpha) \end{cases} , \tag{4.5}$$

which is illustrated by Figure 1.

As a numerical example, we set $\mu = 10\%$, $\sigma = 20\%$, $r = 2\%$, $\alpha = 10\%$ and calculate $\hat{T} = 18$ [years]: A planning horizon T less than 18 years entails only risk-free investment, whereas $T > 18$ has the consequence that the initial capital $v(0)$ must be entirely invested into the stock market.

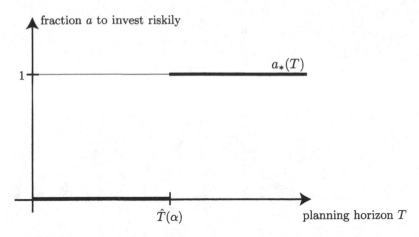

Figure 1. The optimal fraction $a_*(T)$ is no longer a corner solution. However, it is piecewise composed of corner solutions

Kataoka combined with regret

To begin with we identify the "unfavorable extremal events" with both tails of the return distribution and truncate in a symmetrical way. Then only $R(T)$−values

$$\underline{q} \leq R(T) \leq \overline{q} \tag{4.6}$$

are relevant, where $\underline{q} = \frac{\alpha}{2}$−quantile, $\overline{q} = (1 - \frac{\alpha}{2})$−quantile of $R(T)$. In terms of the $\frac{\alpha}{2}$−quantile of $N(0,1)$ we have

$$
\begin{aligned}
\underline{q} &= \left(\mu - \frac{\sigma^2}{2}\right)T + z(\tfrac{\alpha}{2})\sigma\sqrt{T} \\
\overline{q} &= \left(\mu - \frac{\sigma^2}{2}\right)T - z(\tfrac{\alpha}{2})\sigma\sqrt{T}.
\end{aligned}
$$

For planning horizons $T < \hat{T}(\frac{\alpha}{2})$ the relative position of $\underline{q}, rT, \overline{q}$ is as follows

$$\underline{q} < 0 < rT < \overline{q}$$

and is illustrated by Figure 2.

The maximal regret corresponds to the discrepancy between rT and $\underline{q}$ or $\overline{q} - rT$, whichever is bigger. Formally, the maximal regret related to action a is (compare (3.4)):

$$\max \text{regret}(a) = v(0) \max \left\{(1 - a)\left[e^{\overline{q}} - e^{rT}\right] ; a\left[e^{rT} - e^{\underline{q}}\right]\right\}.$$

The intersection of the two straight lines in the curved brackets gives the solution $a_*(T)$.

Both Figures 2 and 3 refer to the case $T < T(\frac{\alpha}{2})$. Otherwise $\underline{q}$ and $\overline{q}$ are bigger than rT and the upwards sloped straight line in Figure 3 gets a negative slope; the optimal

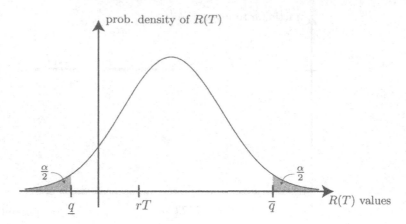

Figure 2. Only return values in the interval $[\underline{q}, \overline{q}]$ are relevant for the planning procedure

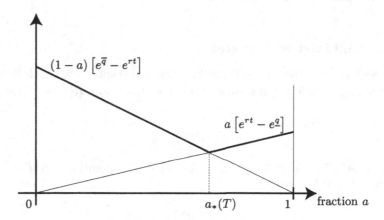

Figure 3. The kinked solid line corresponds to the maximal regret, and the minimum defines the optimal fraction $a_*(T)$

fraction turns out to be $a_* = 1$. Taken together, the optimal fraction is given by an explicit formula, namely

$$
a_*(T) = \begin{cases} \dfrac{\exp\left\{\left(\mu-\frac{\sigma^2}{2}\right)T-z\sigma\sqrt{t}\right\}-\exp\{rT\}}{\exp\left\{\left(\mu-\frac{\sigma^2}{2}\right)T-z\sigma\sqrt{T}\right\}-\exp\left\{\left(\mu-\frac{\sigma^2}{2}\right)T+z\sigma\sqrt{T}\right\}} & , \quad \text{if} \quad T < \hat{T}(\frac{\alpha}{2}) \\ 1 & , \quad \text{if} \quad T \geq \hat{T}(\frac{\alpha}{2}) \end{cases} \tag{4.7}
$$

where z is now the $\frac{\alpha}{2}$–quantile of $N(0,1)$. Graphically, we get a curve which increases with T (cf. Figure 4).

The property $a_*(T) \to 0.5$ $(T \to 0)$ stems from Bernoulli l'Hospitals's rule. Of course, the property is not very important for a long-term investor. Nevertheless, it should be

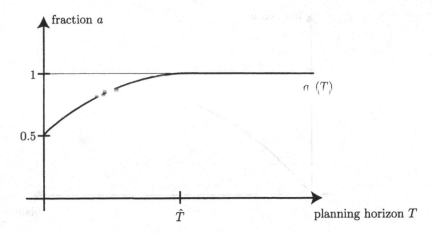

Figure 4. Optimal fraction $a_*(T)$ over planning horizon T

noted that this property also holds in the case of bear markets $(\mu - \frac{\sigma^2}{2} - r \leq 0)$. To see whether this rather robust 50:50 allocation is due to the symmetry of the normal distribution or to the symmetric truncation of both tails we will now turn to the one-sided truncation. The derivation of the optimal fraction $a_*(T)$ along the lines leading to formula (4.7) yields:

- If the investor is rather optimistic and truncates the lower tail we get

$$a_*(T) = 1 \quad \text{for all} \quad T.$$

- If the investor is not too optimistic (perhaps the most plausible case) it is appropriate to truncate the upper tail, thus eliminating astronomically high returns from the planning procedure. Under ordinary conditions (i.e. $\mu - \frac{\sigma^2}{2} - r \geq 0$) the optimal fraction $a_*(T)$ is

$$a_*(T) = 1 - \exp\left\{-\left(\mu - \frac{\sigma^2}{2} - r\right)T - z(\alpha)\sigma\sqrt{T}\right\}. \tag{4.8}$$

The curve in Figure 5 looks like the cumulative distribution function of a positive random variable. Actually, $\mu - \frac{\sigma^2}{2} - r = 0$ leads to the cdf of a Weibull distribution.

Hence the property $a_*(0) = 0.5$ of (4.7) is a consequence of the symmetric truncation. The fact that the curve (4.8) starts from the origin $(a_*(0) = 0)$ fits better to common sense and backs up the right-hand truncation.

5 Concluding remarks

Section 4 focussed on the main cases. Of course, other asymmetric kinds of truncation are conceivable. Furthermore, there exist other regret or disappointment theories which strongly differ from the original Savage/Niehans approach underlying this paper, compare

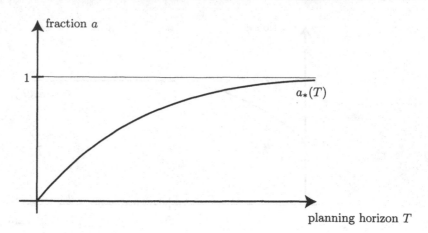

Figure 5. The optimal fraction $a_*(T)$ is a concave, increasing function which approaches 1 asymptotically

for instance Bell (1982) and Bell (1985), Loomes and Sudgen (1982), Loomes and Sudgen (1986), Gul (1991). Finally, Section 4 discussed the case of a regular or bull market (i.e. a nonnegative risk premium $\mu - \frac{\sigma^2}{2} - r$). Bear markets ($\mu - \frac{\sigma^2}{2} - r < 0$) are also possible, at least temporarily. Formulas (4.7) and (4.8) for the optimal fraction $a_*(T)$ are still valid, but the curves look dissimilar to those of Figures 4 and 5. Instead of Figure 4 (symmetrical truncation) we have a convex and downward sloping curve, starting from $a_*(0) = 0.5$ and intersecting the abscissa at $\hat{T}(\frac{\alpha}{2})$; for $T \geq \hat{T}(\frac{\alpha}{2})$ we have $a_*(T) = 0$. Instead of Figure 5 we have a concave curve starting from the origin and intersecting the abscissa at $T = \hat{T}(\alpha)$; for $T \geq \hat{T}(\alpha)$ we have again $a_*(T) = 0$. Thus in bear markets the optimal fraction (according to the Kataoka minmax regret criterion) is in the positive territory only for a very short planning horizon T, then turns into a tendency to shortselling of stocks. Since the model restricts a to $[0; 1]$ the optimal fraction cannot be negative and remains zero.

To sum up: The conjecture is unfounded that outside the expected utility frame work mainly implausible and crude corner solutions are prevalent. However, the original versions of the Wald or regret criterion have to be combined with Kataoka in a suitable manner. The resulting optimal fraction $a_*(T)$ can be represented through an explicit formula, for instance formula (4.8). The ingredients are easier to assess than a von Neumann/Morgenstern utility function. Nevertheless, if one intends to implement (4.8) for practical financial consulting one should check whether the investor is sensitive with respect to regret, is willing or able to decide about the amount $v(0)$, the planning horizon T, the probability α (as rough proxy for risk tolerance) and the parameters μ, σ, r of the decision field.

Bibliography

G. Bamberg, G. Dorfleitner, and R. Lasch, Does the Planning Horizon Affect the Portfolio Structure? In W. Gaul and H. Locarek-Junge, editors, *Classification in the Information Age*, Springer-Verlag, Berlin et al., 100 114, 1000

G. Bamberg and G. Dorfleitner, Is Traditional Capital Market Theory Consistent with Fat-Tailed Log Returns? *Zeitschrift für Betriebswirtschaft* 72:865-878, 2002.

D. Bell, Regret in Decision Making Under Uncertainty, *Operations Research* 30:961-981, 1982.

D. Bell, Disappointment in Decision Making Under Uncertainty, *Operations Research* 33:1-27, 1985.

F. Gul, A Theory of Disappointment Aversion, *Econometrica* 59:667-686, 1991.

S. Kataoka, A Stochastic Programming Model, *Econometrica* 31:181-196, 1963.

W. Krämer and R. Runde, Stochastic Properties of German Stock Returns, *Empirical Economics* 21:281-306, 1996.

A. Lo and C. MacKinlay, Stumbling Block for the Random Walk, In Financial Times, editor, *Mastering Finance*, London, 185-191, 1998.

G. Loomes and R. Sugden, Regret Theory: An Alternative Theory of Rational Choice Under Uncertainty, *Economic Journal* 92:805-824, 1982.

G. Loomes and R. Sugden. Disappointment and Dynamic Consistency in Choice Under Uncertainty, *Review of Economic Studies* 53:271-282, 1986.

S.T. Rachev and S. Mittnik, *Stable Paretian Models in Finance*, Chichester et al., 2000.

H.-J. Wolter, Shortfall-Risiko und Zeithorizonteffekte, *Finanzmarkt und Portfolio Management* 7:330-338, 2000.

H. Zimmermann, Zeithorizont, Risiko und Performance: Eine Übersicht, *Finanzmarkt und Portfolio Management* 5:164-181, 1991.